LEARN ROBOTICS WITH
RASPBERRY PI®

LEARN ROBOTICS WITH RASPBERRY PI®

BUILD AND CODE YOUR OWN MOVING, SENSING, THINKING ROBOTS

BY MATT TIMMONS-BROWN

no starch
press

SAN FRANCISCO

Printed in USA

Third printing

24 23 22 21 20   3 4 5 6 7 8 9

ISBN-10: 1-59327-920-5
ISBN-13: 978-1-59327-920-2

Publisher: William Pollock
Production Editor: Janelle Ludowise
Cover Illustration: Josh Ellingson
Cover Design: Mimi Heft
Interior Design: Beth Middleworth
Developmental Editor: Liz Chadwick
Technical Reviewer: Jim Darby
Copyeditor: Rachel Monaghan
Compositor: Kim Scott, Bumpy Design
Proofreader: James Fraleigh

Circuit diagrams made using Fritzing (*http://fritzing.org/*).

The following images are reproduced with permission:

Figure 6-2 was created by László Németh. Figure 6-3 was created by Phillip Burgess and is licensed under the Creative Commons Attribution-ShareAlike 3.0 Unported License (*https://learn.adafruit.com/assets/10668*). Figure 7-17 was created by AlexJ. Figure 8-19 was created by SharkD and is licensed under the Creative Commons Attribution-ShareAlike 3.0 Unported License (*https://commons.wikimedia.org/wiki/File:HSV_color_solid_cylinder_saturation_gray.png*).

For information on distribution, translations, or bulk sales, please contact No Starch Press, Inc. directly:

No Starch Press, Inc.
245 8th Street, San Francisco, CA 94103
phone: 1.415.863.9900; info@nostarch.com
www.nostarch.com

Library of Congress Cataloging-in-Publication Data:
Names: Timmons-Brown, Matt, author.
Title: Learn robotics with Raspberry Pi : build and code your own moving,
    sensing, thinking robots / Matt Timmons-Brown.
Description: San Francisco : No Starch Press,Inc., [2019]
Identifiers: LCCN 2018042503 (print) | LCCN 2018048396 (ebook) | ISBN 9781593279202 (print)
    | ISBN 1593279205 (print) | ISBN 9781593279219 (ebook) | ISBN 1593279213
    (ebook)
Subjects: LCSH: Robotics. | Raspberry Pi (Computer)
Classification: LCC TJ211 (ebook) | LCC TJ211 .T579 2019 (print) | DDC
    629.8/9--dc23
LC record available at https://lccn.loc.gov/2018042503

TO MY PARENTS, REBECCA AND JEFF, FOR
SUPPORTING YOUR SON'S STRANGE PASSION
FOR PI AND ALWAYS BELIEVING IN ME.
THANK YOU FOR YOUR INVALUABLE ADVICE
AND UNLIMITED LOVE.

I DEDICATE THIS BOOK TO YOU.

CONTENTS

CONTENTS IN DETAIL

ACKNOWLEDGMENTS

This publication would not have been possible without the help, support, and hard work of a plethora of different people—the full list of which would probably span the length of this entire book.

Firstly, I would like to thank Liz Chadwick, Janelle Ludowise, Bill Pollock, and the entire team at No Starch Press for persuading me to write a book in the first place and for making *Learn Robotics with Raspberry Pi* a reality. Thank you also to Jim Darby, Raspberry Pi extraordinaire and friend, for diligently turning his hands to the technical review.

Without the Raspberry Pi computer, not only would this book not exist, but neither would my fascination and love for computer science as I know it today. For this, I have to thank the Raspberry Pi Foundation and everybody involved with it. The impact of the Pi and the Foundation's work has changed and improved the world in so many ways. In particular, I am indebted to Eben Upton for his years of advice, mentorship, and support—and for writing the foreword to this book too.

Raspberry Pi would be nothing without the immense, worldwide community that surrounds it. Regardless of whether you're brand new to this community or have been in it since the start, thank you . . . And extra thanks to all those that have supported my YouTube channel, come to any of my talks, or smiled at me at Pi events!

Thanks to Phil Howard, Ben Nuttall, and Simon Beal for their Python 3 support and expertise throughout the writing process. To Paul Freakley, Brian Corteil, and Rob Karpinski—thank you for helping me become a maker and giving me access to crazy laser cutters, 3D printers, and more.

Thank you to Tim Hanbury-Tracy for giving me your place in the pre-order queue for the original Raspberry Pi back in 2012—how different my life would have been if not for your generosity.

Finally, thank you to my friends and family; without you my journey in computer science and the writing of this book would not be possible. In particular, thank you to my parents, Rebecca and Jeff, for your unending love, support, and guidance.

FOREWORD

When we founded the Raspberry Pi Foundation in 2008 and set out to develop the Raspberry Pi computer, we saw it primarily as a platform for software development. If you had asked me then what our predominant educational use case would be in 2018, I would probably have cited game development: after all, that had been my route into computing back in the late 1980s.

In the six years since we launched the first Raspberry Pi, the community around our little educational computer has grown beyond our wildest dreams. We've seen children and adults all over the world using Raspberry Pi to learn engineering skills; we've sent two units to the International Space Station, where they have run code developed by over 3,000 teams of young people from across Europe; and we've trained thousands of educators to deliver compelling lessons using our library of free educational resources.

All this has been a surprise, but the biggest surprise for me has been the popularity of *physical computing* projects: not just writing code, but using it to sense, control, and interact with the real world. While moving sprites around on the screen is cool, moving physical objects around the room is cooler. The 40-pin GPIO connector, included at my colleague Pete Lomas's insistence, has proven in many ways to be the single most useful feature on the board.

It's a truism that platforms are only as good as their documentation, and for many beginners, the learning curve for physical computing can appear daunting. *Learn Robotics with Raspberry Pi* provides a gentle introduction to this exciting field, building up from the simplest input and output examples to a robot which incorporates wireless control and is capable of autonomously tracking and following lines and objects.

My hope is that in twenty or thirty years, a handful of people will look back on Raspberry Pi with the same affection I have for the BBC Micro and Commodore Amiga. If that happens, I'm sure some of those people will have *Learn Robotics with Raspberry Pi* to thank for showing them how to get the most out of the platform.

Dr. Eben Upton, CBE FREng
CEO, Raspberry Pi (Trading), Ltd.
Cambridge, UK
April 2018

INTRODUCTION

WELCOME TO *LEARN ROBOTICS WITH RASPBERRY PI*! IN THIS BOOK, YOU AND I WILL GO ON AN EXCITING ADVENTURE THROUGH ELECTRONICS, CODE, AND ROBOTICS. I'LL SHOW YOU HOW TO USE THE RASPBERRY PI MINICOMPUTER TO BUILD YOUR VERY OWN CUSTOMIZABLE ROBOT FROM THE GROUND UP.

Along the way, we'll undertake a series of projects that will give your robot awesome abilities, from remote control to following lines to recognizing objects and more!

By the end of this book you'll be equipped with the programming and engineering skills to embark upon years of robotics and computer science fun, and should have the basic understanding you need to make your craziest robot ideas come to life.

This book also introduces many other areas of computing along the way, including coding in one of the most popular programming languages around: Python. This is the perfect first step for anybody with an interest in computers and technology!

WHY BUILD AND LEARN ABOUT ROBOTS?

Robots are all around us. They build the products you use every day. They save lives in surgery. They even explore Mars and the rest of our solar system. As tech improves, humans rely on robots more and more to make our lives easier, better, and safer. With the rise of artificial intelligence, it won't be long until things like driverless cars and smart robotic companions are the norm!

There has never been a better time to learn about robotics, whether that's just to satisfy your own curiosity or to embark on a future lucrative career. And besides, if you understand even just a *little* bit about robots, you'll have a much better chance at surviving the robot uprising. (Just kidding.)

By building robots you gain experience and understanding in a vast range of areas. Making things? *Check!* Electronics engineering? *Check!* Software programming? *Check!* This book is the perfect introduction to all three.

But beyond all that, I can boil down why you should build robots into a single statement: **because it's fun**. There is something uniquely intriguing and exciting about watching something *you* have made run around on the floor, avoid obstacles, or flash lights.

Robotics hooked me into the world of computer science when I was 13 years old and it hasn't let go since.

WHY THE RASPBERRY PI?

The Raspberry Pi is a $35 credit card–size computer that was created to provide people with an inexpensive introduction to programming and electronics. Despite its small size and low cost, each Pi is a fully functional computer that can do pretty much everything you would expect, from running programs to word processing and web browsing.

Raspberry Pi is a great platform to learn robotics with. It's cheap, small, easily powered, and incredibly accessible. You can program a Pi in almost any language you can imagine and embed it into all sorts of electronics projects. A Pi hits just the right sweet spot between power and simplicity, meaning you can create robots without limitations.

Since it was launched out of Cambridge, UK, in 2012, the Raspberry Pi has gained a massive worldwide following and community. Millions of people share their progress, projects, and ideas online, which makes it easy for beginners to improve and learn from someone who knows more. There are also many in-person events where you can chat and show off your achievements. These are usually lovingly called *Raspberry Jams*, and they take place around the globe.

WHAT IS IN THIS BOOK?

This book is project-based, and revolves around a two-wheeled robot that I'll show you how to make from scratch. You'll improve this robot project by project, adding components and coding new functionality. At each stage of the book I'll provide comprehensive instructions and explanations of each build and the program behind it. You can also download the full code and resources for free from *https://nostarch.com/raspirobots/*. Any updates or further notes from the book can also be found there.

Take a look at what's in store for you in each chapter:

Chapter 1: Getting Up and Running takes you on a tour of the Raspberry Pi and its features. I also show you how to install the operating system and set it up for use over your local network using SSH. In this chapter you'll meet the terminal and write your first Python program.

Chapter 2: Electronics Basics introduces electricity, what it is, and how we can harness it. You'll find two beginner projects in this chapter that are great for starting off your adventures before you start making robots. By the end, you'll be able to make circuits that blink an LED and respond to a button.

Chapter 3: Building Your Robot begins your robotic journey. Here you'll start to build your robot! We'll build the base, with motors and wheels, and you'll find lots of guidance about the different parts of your robot and how to wire it up.

Chapter 4: Making Your Robot Move gives your fully constructed robot the power of movement, with complete remote control using a Nintendo Wiimote. You'll use Python code to make your robot move, first in a simple pattern and then by just tilting and orienting a Nintendo Wiimote, *Mario Kart*–style.

Chapter 5: Avoiding Obstacles gives you your first taste of robotic autonomy. In this project you'll use an ultrasonic distance sensor to give your robot the ability to sense and avoid obstacles in its way. You'll never crash again!

Chapter 6: Customizing with Lights and Sound allows you to customize your robot with super-bright lights and speakers. You'll be able to program your own dazzling light shows, and connect a 3.5 mm speaker to your Raspberry Pi so that your robot can emit noises, like a car horn.

Chapter 7: Line Following shows you how to use sensors and code to make your robot follow a black line. It will be racing around a track all by itself in no time at all!

Chapter 8: Computer Vision: Follow a Colored Ball is the most advanced project of this book and introduces image processing, one of the most high-tech areas of computer science. In this chapter your robot will use the official Raspberry Pi Camera Module and computer vision algorithms to recognize and follow a colored ball, no matter where it is in your robot's environment.

WHO IS THIS BOOK FOR?

Learn Robotics with Raspberry Pi is a book for anybody who is interested in robots, programming, and electronics. No assumptions are made about ability, and I steer well clear of confusing, unexplained jargon throughout. People of all ages and backgrounds can learn something from the easy-to-follow projects and guidance.

WHERE SHOULD YOU BUY PARTS?

Over the course of this book and the projects it contains, you'll need various bits and pieces: electronic components, making materials, and a few more tools. Don't worry, though—everything is affordable and widely available. There will be specific advice in each chapter, but in general everything you need can be picked up online on sites such as eBay and Amazon.

All of the electronic components in the book can be purchased from online shops like eBay (*https://www.ebay.com/*) or dedicated online electronics stores such as Adafruit (*https://www.adafruit.com/*), Pimoroni (*https://shop.pimoroni.com/*), The Pi Hut (*https://thepihut .com/*), CPC Farnell (*http://cpc.farnell.com/*), and RS Components (*http://uk.rs-online.com/web/*). This list is by no means exhaustive and you may find cheaper, closer online retailers in your own country. You might even be fortunate enough to have a local electronics hardware store where you can grab your stuff!

I'll introduce and explain the exact parts you'll need in each project, but here's a full list of everything you'll use in the book:

Chapter 1
- Raspberry Pi 3 Model B+
- 8GB+ microSD card
- HDMI cable, USB keyboard/mouse
- 5 V micro USB power adapter

Chapter 2
- 400-point breadboard
- An LED with appropriate resistor
- M-F/F-F/M-M jumper wires
- Momentary push button

Chapter 3

- A chassis for your robot (I make mine out of LEGO)
- Two brushed 5 V to 9 V motors with tires
- Six AA battery holder
- Six AA batteries (I recommend rechargeable)
- LM2596 buck converter module
- L293D motor driver chip

Chapter 4

- Nintendo Wii remote
- **For earlier models:** Bluetooth dongle for Pis prior to Model 3/Zero W

Chapter 5

- HC-SR04 ultrasonic distance sensor
- A 1 kΩ resistor and a 2 kΩ resistor

Chapter 6

- NeoPixel stick with headers
- A small 3.5 mm speaker

Chapter 7

- Two TCRT5000-based infrared line-following sensor modules

Chapter 8

- An Official Raspberry Pi Camera Module
- A colored ball

You'll also find these tools/materials handy along the way:

- Variety of screwdrivers
- Hot-glue gun
- Multimeter
- Soldering iron
- Wire stripper
- Sticky tack/Velcro/3M Dual Lock

LET'S GET STARTED!

In short: robots and Raspberry Pi are awesome. Now that that's out of the way and we're acquainted, it's time to get going! Just turn the page to begin your robotic adventures . . .

1
GETTING UP AND RUNNING

THIS BOOK WILL TEACH YOU HOW TO BUILD YOUR VERY OWN ROBOTS. THIS EXCITING ADVENTURE WILL INVOLVE ALL SORTS OF ELECTRONICS, MAKING, AND PROGRAMMING.

Over the following chapters I'll guide you through everything you need to know how to do, from connecting LEDs, buttons, batteries, and motors to getting your robot to follow lines, giving it sensors to see the world, and more! You'll use a Raspberry Pi in all the projects to come, so now that you know what one is, let's get your Pi set up.

GET YOUR HANDS ON A RASPBERRY PI

Before you can proceed, you'll need a Raspberry Pi, of course! The Raspberry Pi is available worldwide, so it should be easy to buy one, no matter where you live.

At the time of writing, there are several different Raspberry Pi models available. The three most up-to-date ones are the *Raspberry Pi 4*, the *Raspberry Pi 3 Model B+*, and the *Raspberry Pi Zero*. The Raspberry Pi 3 Model B+ is the updated classic $35 Raspberry Pi, shown in Figure 1-1. This is the board I'll be using throughout this book, as it's the best option for development: it has more full-size connectors. This saves you from fiddling around with the adapters and USB hubs that you'd need for the Pi Zero.

FIGURE 1-1

The Raspberry Pi 3
Model B+

The Raspberry Pi Zero is a stripped-down, smaller board that retails for just $5. The Pi Zero W is identical except that it's the *wireless* version, meaning it's equipped with Wi-Fi and Bluetooth capabilities, and retails for $10. The Zero and Zero W are shown in Figure 1-2.

You might be wondering why we're not using the smaller, more compact Pi Zero. After all, a smaller board would take up less space so you could make a robot smaller or have space for more hardware.

But if you're using the Pi Zero, you'll need USB and HDMI adapters for the miniature ports to plug in USB devices or a monitor, which is more challenging. These adapters have to be purchased separately too. When you have some more experience in the field of robotics, you may decide that it's worth using a Pi Zero for one of your later projects; when that time comes, go for it!

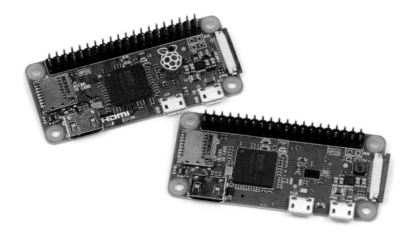

FIGURE 1-2

The Raspberry Pi Zero (left) and Raspberry Pi Zero W (right)

If you already have an older model of the Pi, don't worry. The version of Raspberry Pi you have doesn't really matter, since all of them are cross compatible, and you can use any of them to build the robots in this book. The only difference is that you'll need adapters for things like your wireless internet connection. Table 1-1 gives the different specs for each model.

MODEL	Raspberry Pi 3 Model B+	Raspberry Pi Zero	Raspberry Pi Zero W
RAM	1GB	512MB	512MB
PROCESSOR	64-bit quad-core 1.4GHz	32-bit single-core 1GHz	32-bit single-core 1GHz
PORTS	HDMI, 4x USB 2.0, Micro-USB power	Mini-HDMI, Micro-USB (data), Micro-USB power	Mini-HDMI, Micro-USB (data), Micro-USB power
CONNECTIVITY	Wi-Fi, Bluetooth, Ethernet	None	Wi-Fi, Bluetooth
PRICE	$35	$5	$10

TABLE 1-1

Main Raspberry Pi Model Specification Comparison

Recently, the Raspberry Pi 4 has come out. While this is fully compatible with this book, it uses more power than the 3B+ and will drain your robot's batteries faster, so I recommend sticking with the Pi 3B+. At the end of the day, don't worry about which Pi you have— you'll be able to follow along regardless.

You can find a distributor for your country on the Raspberry Pi Foundation website (*https://www.raspberrypi.org/products/*).

Your First Taste of Pi

When you see your Raspberry Pi for the first time, you might find yourself bewildered. You probably associate a normal "computer" with a screen, keyboard, mouse, and storage—but the Raspberry Pi is a little different.

Unbox it, and you'll find a bare-looking board with all sorts of components sticking out of it. If you have a Pi 3 B+, it should look exactly like Figure 1-3. Later versions might look slightly different, but they all have the same basic features.

NOTE

Raspberry Pi is not the only platform or computer you can build robots with, but it is one of the easiest and most accessible ways of doing so!

FIGURE 1-3

A top view of the Raspberry Pi 3 Model B+

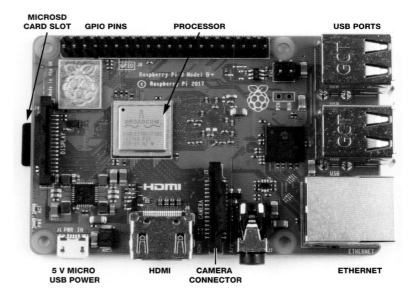

Let's walk through what all these components do:

USB ports   There are four USB ports you can use to plug in USB keyboards, mice, USB sticks, and other devices.

Ethernet network port   This is for a wired internet connection.

HDMI port   *HDMI* stands for *High-Definition Multimedia Interface*, and this port is what you'll use to connect your Pi to a screen like a TV or a computer monitor.

Micro USB power jack   This is where you'll plug in the 5 V of power that every Raspberry Pi requires to work; this is the same sort of power input as many mobile phones. It's also worth noting that there is no power button! Your Pi will be on for as long as you keep the power cable connected.

MicroSD card slot   While most computers have some form of built-in storage—your laptop often has a hard drive, for example—a Raspberry Pi has no "onboard" storage. Instead, the software your computer runs on, known as the *operating system* (*OS*), and all of your files are stored on a microSD card, much like you might find in a digital camera. As part of the setup process, I'll show you how to configure a microSD card and install the OS you need for your Pi.

Quad-Core 1.4GHz processor   In the middle of your Pi you'll see the brain of your computer. You may be wondering just how powerful your new purchase is: is it as fast as a laptop or a desktop computer? The processor, coupled with the Pi's 1GB of RAM, gives the Raspberry Pi power that's roughly equivalent to some smartphones. As you'll find out later, you can do a *serious* amount of computing with this processor.

Camera connector   Next to the HDMI port is a clip-like connector labeled *camera*. It's the input for the official Raspberry Pi Camera Module—something you'll use in this book to give your robot the ability to see!

GPIO pins   One of the most alien features of a Raspberry Pi are the 40 shiny metal pins found on the top edge of the board, shown in Figure 1-4. These are called *GPIO pins*, which stands for General-Purpose Input/Output. You can program these pins to control a huge variety of electronics, components, and other parts including LEDs, sensors, and motors (see Figure 1-5).

FIGURE 1-4

The GPIO pins

FIGURE 1-5

A selection of hardware
that you can connect to
your Raspberry Pi using
the GPIO pins

These GPIO pins are the gateway to a world of *physical comput-ing*. You'll be using them to wire up the electronic elements of your robot (motors, line-following sensors, and more). I'll then guide you through programming this new hardware so that it does your bidding!

What You'll Need

As you've noticed, your new computer is lacking in some funda-mental features, like a screen and a keyboard. You'll need some extra hardware to get it set up and running. Fortunately, you'll likely have most of these lying around already.

A 5 V Micro-USB power adapter   This is used to power the Raspberry Pi. Any old Android smartphone charger should be fine. Most power adapters will list their output voltage and current, so you just need to ensure the output voltage of the charger is 5 V and the output current is at least 2.5 A. There are a lot of power adapters around that *don't* meet this specification. It is quite common to encounter mysterious faults because of one of these supplies. If you don't have a spare power adapter lying around, grab an official one here: *https://www.raspberrypi.org/products/raspberry-pi-universal-power-supply/*.

A USB keyboard and mouse   By default the Raspberry Pi has no form of input, so you'll need both a USB keyboard and mouse in order to interface with it in the beginning. If you have a desktop PC at home, just yank out the existing USB keyboard and mouse and use those. If not, you can pick them up online or at any computer store.

An 8GB+ microSD card   As mentioned, the Raspberry Pi has no onboard storage, so you'll need a microSD card (or a normal SD card if you have a first-generation Raspberry Pi) to store the OS. These can also be picked up online or in a computer store. You'll need at least an 8GB card—the more space, the better!

An HDMI cable   This will be used to connect your Raspberry Pi to an HDMI TV or monitor. This is a standard cable that you can pick up online or in your local store.

A monitor or TV   You will need some sort of display output for your Raspberry Pi. Anything with an HDMI port will do, whether it's a computer monitor, TV, or another type of screen. Many computer monitors have a DVI input, and you can pick up an HDMI-to-DVI adapter or even cable.

It would also be incredibly helpful if you have access to a desktop computer or laptop. This isn't a *necessity*, but it will be an advantage in a variety of ways. First, you'll need to prepare an SD card with the software your Raspberry Pi will run on, which needs to be done on another machine. Second, you'll be wirelessly connecting to your Pi and controlling it over your local area network. This saves you from having to keep plugging and unplugging your Pi to your monitor, and you'll need a separate computer for that too. By using your Pi over your local area network, you'll only need a monitor or TV for the initial setup process in this chapter. This shouldn't take more than half an hour!

If you aren't able to access another computer, don't fret. You can work around this and still follow along just fine.

In later chapters of this book we'll be using more hardware, components, and electronics, but you don't have to worry about that just yet. I'll be sure to tell you everything you need before we launch into each project.

SETTING UP YOUR RASPBERRY PI

Now that you've gathered all the tech, it's time to set up your Raspberry Pi. This can seem like a daunting task for a beginner, but I'll walk you through it. All you have to do is set up your microSD card, hook up your hardware, and then boot up your Pi and configure a few settings.

If you don't have access to another computer to follow these next steps, you can purchase microSD cards that are preloaded with the OS already set up. You can find these online by searching for "preinstalled NOOBS Raspberry Pi microSD cards."

If you do have another computer available, though, I would recommend installing the OS yourself, as it's a handy skill to know. That way, if anything goes wrong and you need to start afresh, you'll know what to do. Preinstalled microSD cards are also expensive!

Installing Your Pi's Operating System on Windows/macOS

The operating system is the software that every modern computer runs on, and while different operating systems can look quite similar, they're not all the same. You'll likely be most familiar with Windows or macOS, but your Raspberry Pi runs Linux operating systems.

Linux is a family of free and open source operating systems with different *distributions*, meaning that there are different variations of Linux for different purposes. For Raspberry Pi, most people use the *Raspbian* distribution, the operating system officially supported by the Raspberry Pi Foundation (see Figure 1-6). Raspbian was developed and refined to run smooth as butter on your Pi, and you'll find it has many features in common with the OS you normally use.

FIGURE 1-6

The Raspbian desktop environment

Preparing Your SD Card

Before you can install Raspbian on your microSD card, you first have to clear out anything that might already be stored on it. Even if your card is brand new, I recommend doing this, because it can some-times come with stuff already on it. This process is called *formatting* your microSD card. Be sure to read the warning in the sidebar before you format your microSD card!

1. Insert your microSD card into your normal computer. Some computers have SD card or microSD card ports, but many don't. If your computer doesn't have a place you can plug in your SD card, you'll need to use a *USB SD card adapter*, like the one shown in Figure 1-7. This small device lets you plug your card into one of the USB slots on your PC. You can find this easily and cheaply online (just search "SD card USB adapter") or in your local computer store.

WARNING

During the formatting pro-cess, the storage device you have selected will be entirely and irreversibly erased. Make sure you double-check that you have selected the right drive name so you don't accidentally delete every-thing from another device and lose your data.

2. Once you have your SD card plugged in, you should be able to navigate to it in your file explorer. If you're using Windows, look under Devices and Drives; if you're using a Mac, locate it using Finder. Make a note of the drive name of your microSD card. This is the letter that your computer assigns to it when it's plugged in (like *D:* or *H:*).

3. The best way to ensure your SD card is completely wiped and formatted correctly is to use the official formatting software *SD Card Formatter*. To install this software, visit *https://www.sdcard .org/downloads/*, click **SD Memory Card Formatter** in the menu, and find the formatter for your operating system. It will ask you to accept the conditions, so scroll to the bottom and click **Accept** to continue, and the latest version of the software should download. After this download has finished, run the installation setup, which looks like Figure 1-8. Make sure to follow the instructions and accept the terms and conditions.

FIGURE 1-8

The SD Card Formatter installation process

4. Once this installation process has finished, find and run the SD Card Formatter. A window like the one in Figure 1-9 will open. The process from here is very simple: select your card from the drop-down menu (remember the drive name you noted earlier), keep the option **Quick Format** selected, and then click **Format**. Watch the progress bar as your card is successfully formatted!

Installing NOOBS

Now you have a blank, formatted microSD card, so it's time to install the Raspbian OS onto it. This process has been made super simple with *NOOBS*—New Out Of the Box Software—from the Raspberry Pi Foundation. Just follow these instructions:

1. With your microSD card plugged into your computer, navigate to the Raspberry Pi website at *https://www.raspberrypi.org/*. Click **Downloads** at the top of the page, and from there click the **NOOBS** link. Make sure you download the latest *full* version of NOOBS—don't install the Lite version, as it won't allow you to set up your Raspberry Pi unless it's connected to the internet. Find the button labeled **Download ZIP** and click it. Wait for your operating system to download. This may take anywhere from a few minutes to a few hours, depending on internet speed.

2. Once NOOBS has finished downloading, find it in your *Downloads* folder and extract the compressed file. In Windows, right-click the file and select **Extract All**, choose a place on your computer

to store your extracted operating system files, and click **Extract**. On Mac, downloading through Safari will result in a NOOBS file that is automatically extracted for you.

3. Finally, navigate to your newly downloaded NOOBS files and copy and paste them onto your microSD card. You can do this by highlighting them all with your mouse, copying them, and then pasting them onto your formatted card, as shown in Figure 1-10.

FIGURE 1-10

Transferring all of the extracted files to the microSD card

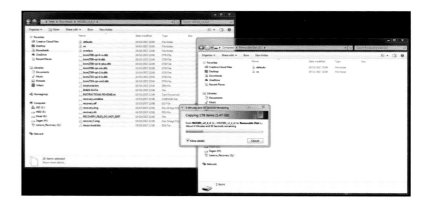

Huzzah! You have successfully set up your SD card and can safely remove it from your computer. Make sure to right-click on the device and then eject the card before you remove it.

Plugging In and Wiring Up Your Raspberry Pi

The next stage is physically setting up your Raspberry Pi. You'll want to do this in an area that has some space, access to a Wi-Fi network, and access to your monitor, whether that's a TV or computer screen.

1. Unbox your Raspberry Pi and plug your microSD card into the slot at the bottom of the board (see Figure 1-3 for a reminder of where this is). Make sure it's pushed all the way in. Some models of Pi (the 1 B+ and 2 B) make a "click" sound when the card is inserted properly, but the later 3 B/B+ does *not*.

2. Connect your USB keyboard and mouse to the USB ports on your Raspberry Pi.

3. Plug your HDMI cable into the HDMI port on your Pi. Insert the other end into your TV or monitor.

4. Now, to start up your Raspberry Pi, connect your 5 V Micro-USB power cable to the power input next to the HDMI port (see Figure 1-11). You should see a flurry of LEDs and some activity on your screen. Congratulations! You have brought your Pi to life!

If nothing has come up on your screen, try flicking between the HDMI inputs with your remote to find the right one—most screens have multiple inputs. As a general rule of thumb, you should turn your monitor on before you plug in power to your Pi; some screens fail to pick up the HDMI signal if this is done the other way around! So try turning the screen off, plugging in your HDMI cable, turning the screen on, and *then* plugging the 5 V micro USB power cable into your Raspberry Pi.

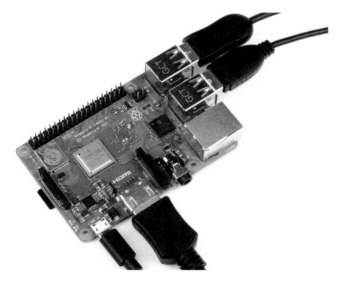

FIGURE 1-11

A happily working Raspberry Pi

If you want a wired internet connection and you're near your internet router, you could hook the Pi up to the internet by plugging an Ethernet cable from your router into the Ethernet port next to the USB ports. However, I'll be using a wireless internet connection throughout this book and I recommend you do so too, to free you up in terms of movement.

If you see a lightning bolt symbol at the top right of the screen, this is the Pi warning you that the power supply isn't providing enough power and the system may be unreliable. In this case you should use a better power supply, ideally the official one.

Installing Raspbian

On the screen connected to your Raspberry Pi, you should see a NOOBS interface like that displayed in Figure 1-12. Now all you need to do is install and configure Raspbian OS. This is super-easy and you'll be up and running in no time!

To install the OS, simply click the box next to the Raspbian option and then click **Install** at the top of the NOOBS box. Raspbian should extract itself and set up your SD card for you. Sit back, relax, and watch the progress!

FIGURE 1-12

The NOOBs installation interface

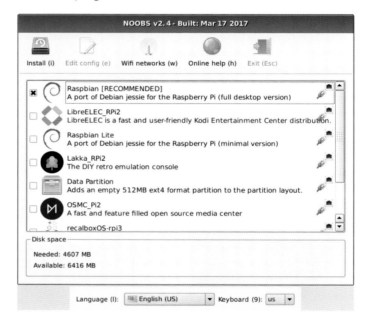

OTHER PI OPERATING SYSTEMS

If you connect the NOOBS interface to the internet, you'll be able to download and access many other Raspberry Pi operating systems. There are lots of other flavors of Linux that you can install, all with their own pros and cons. For example, OSMC (Open Source Media Center) is a video-playing distribution for plugging into a TV as a media center. We'll be sticking to the official Raspbian OS throughout this book. You can return to the NOOBS setup interface at any time by rebooting your Pi and holding the SHIFT key if you want to try out other systems.

Once the installation process has finished, your Pi should reboot into the Raspbian desktop environment, as shown earlier in Figure 1-6.

You will find that a "Welcome to Raspberry Pi" setup wizard starts automatically. As we will be manually configuring your Raspberry Pi's preferences in more detail later, dismiss this wizard by clicking on the **Cancel** option.

You will notice that the Raspbian environment is quite similar to other operating systems you may have used. You will find the menu bar on the top of the screen, as shown in Figure 1-13, with applications on the left side and utilities, such as volume and Wi-Fi, on the right side. Take some time to explore!

FIGURE 1-13
The Raspberry Pi menu

Configuring Raspbian

Before you progress any further, there are two things you should do now to configure Raspbian that will save you time later: change some preferences and set up Wi-Fi.

Changing Raspbian Preferences

When Raspbian is installed, a few important features are turned off by default. This is for security and efficiency reasons, and because lots of people don't need them. You *will* need some of them, however, so we'll enable these now so you don't have to do it later.

In the top-left corner of the screen, click the Pi logo, and from the menu click **Preferences ▸ Raspberry Pi Configuration**. You should see a dialog box like Figure 1-14.

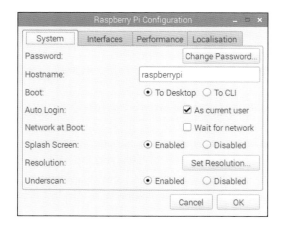

When the application first opens, you'll be greeted by the System tab. Here you can change your account password, screen resolution, and other settings.

By default, your Raspberry Pi will automatically log in. The standard user profile has the username *pi* and the password *raspberry*. I recommend changing the password to something of your choosing, to improve the security of your Raspberry Pi. Make a mental note of the password, as you will need it in the future to log in.

Before you can set up Wi-Fi in the next section, you'll need to tell your Raspberry Pi what country you are operating in. To do this, click the **Localization** tab, then click **Set WiFi Country**. From the drop-down menu, select your current country.

Once you've changed your password, navigate to the next tab, Interfaces. From here, change the following options from Disabled to **Enabled**:

- Camera
- SSH
- VNC
- SPI
- I^{2}C

Enabling Camera will allow you to connect the official Raspberry Pi Camera Module, which we'll do in the last chapter of this book. Enabling SSH and VNC will allow you to remotely access your Pi over your local network, which I'll show you how to do in a second. The other two—SPI and I^{2}C—all relate to functions on the GPIO pins, and we'll come across those later in the book.

With these settings enabled, click **OK** and then reboot your Raspberry Pi when prompted.

Connecting Your Raspberry Pi to the Internet

Connecting your Raspberry Pi to a Wi-Fi network should take only a few minutes. As mentioned, you can connect your Pi to the internet using a physical Ethernet cable—this does not require any configuration. When it comes to building robots, however, you want everything to be as wireless as possible, so I recommend using Wi-Fi.

Also note that to download the software and code we will be using throughout this book, your Pi will need to be connected to the internet.

Both the Raspberry Pi 3 B+ and Pi Zero W have Wi-Fi and Bluetooth built onto the board. If you are using an older model, you'll need to buy a USB Wi-Fi dongle and plug it into one of your Pi's USB ports to connect to the internet. Some Wi-Fi dongles have problems connecting with the Pi, so I recommend using the official model, which you can buy at *https://www.raspberrypi.org/products/raspberry-pi-usb-wifi-dongle/*.

To connect to a Wi-Fi network from the desktop, simply click the Wi-Fi symbol in the top right of the screen and a drop-down list of local networks should appear, as shown in Figure 1-15. Select your network, enter your password, and press **OK**.

FIGURE 1-15

The Wi-Fi drop-down menu

The menu bar icon should then change to show your Wi-Fi strength, indicating your Pi is now connected to the internet.

THE TERRIFIC WORLD OF THE TERMINAL

The *terminal* (or *shell*) is a way for you to give instructions directly for your computer. Specifically, it is an interface in which you can type and execute text-based commands. This is in contrast to the *graphical user interface* (or *GUI*, pronounced "gooey") used by modern operating systems, which you probably know as a desktop. The GUI allows you to use a computer to do many advanced things in simple and visual ways. If you haven't used a Raspberry Pi or a Linux machine before, you most likely will have only ever used a computer through its GUI.

Before computers had the processing power to use complicated graphics, a user operated a computer solely through a terminal using a *command line interface*. This is a purely text-based way of interacting with a computer: you enter text commands to tell the computer what to do, and it responds by simply outputting text.

This may sound like an old-fashioned, not-very-useful way of interacting with computers, but it's still immensely powerful. Using the terminal allows you to perform exact commands efficiently, with only a few characters. The Raspberry Pi terminal is the primary way we will be interacting with our robots.

Touring the Terminal

You'll become more familiar and confident with using the terminal as you work through this book. For now, to acquaint you with it, I'll give you a quick tour and introduce you to how it works.

To open a terminal window from the desktop, click the Raspberry Pi menu in the top left and then click **Accessories ▸ Terminal**. A black box, as shown in Figure 1-16, should appear. You're now using the terminal!

The terminal presents you with a *prompt* that looks something like this:

```
pi@raspberrypi:~ $
```

In this situation, a prompt is simply your Raspberry Pi waiting for you to enter a text command; it is *prompting* you to do something. Unfortunately, the terminal doesn't understand plain English (or any other language, for that matter). Instead, you communicate using specific *commands*. These are predefined phrases and character sequences that tell the terminal how to behave. There are thousands of commands in existence, but don't worry. I'll introduce you to some simple ones now.

You can access files in different folders (known as *directories*) with the terminal the same way you do in your regular OS. The difference is that the terminal interface is like a text-based file manager. At any time you are "inside" a directory, and you can change your location when you want. To explore this, let's practice a few commands.

When you open a terminal, you start out in the Pi's home directory. You can use the first command, ls, to find out what files and folders are in the directory you're currently in. In other words, ls *lists* the contents of your current directory. With your keyboard, enter this command into the terminal:

```
pi@raspberrypi:~ $ ls
```

Press ENTER. You should see a list of the directories and files stored inside your current directory, as shown in Figure 1-17. Now open the same folder in the desktop by going to **Accessories ▶ File Manager**, and it should match what you see in the terminal.

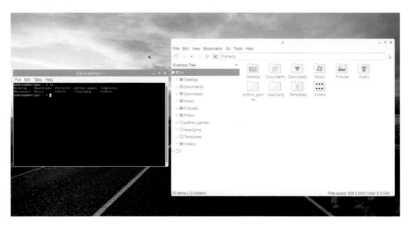

You'll notice that the text is coded in different colors. This helps you quickly identify items. Blue items are directories. That bright green text is your username—in this case, the default user, *pi*.

NOTE

In your Raspberry Pi and Linux adventures, if you're ever unsure about a command or want to know how to use the terminal to do a certain thing, the best way to learn is to look it up online. There's plenty of help and command line manuals to guide you through the process—no matter what you want to do with the terminal!

FIGURE 1-17

The Raspberry Pi terminal side by side with the graphical file manager, displaying the same thing in different ways

You navigate in and out of these folders in text, like you do in a file manager, using the command `cd`, which stands for *change directory*. Pick a directory that was just listed by `ls` and then enter the following:

```
pi@raspberrypi:~ $ cd that_directory
```

Replace *that_directory* with the name of your directory. So, for example, I'll change into the *Documents* directory with the following command:

```
pi@raspberrypi:~ $ cd Documents
```

This moves me into my *Documents* folder. It's important to know that the terminal is *case-sensitive*, meaning that the terminal recognizes upper- and lowercase letters as different things; for example, *d* is not the same as *D*. If I hadn't capitalized the *D* at the start of `Documents` and entered `documents` instead, my Pi wouldn't understand where I wanted to go, and would return an error.

When you move folders, you should notice that the blue text before the $ prompt changes as well, to look something like this:

```
pi@raspberrypi:~/Documents $
```

The terminal will let you know where you are in the filesystem by listing your location before the prompt. This will turn out to be very useful.

In your new folder, enter `ls` again to see what files and folders are inside it. When I do this in *Documents*, I get the output in Listing 1-1.

LISTING 1-1

Looking into my *Documents* folder from the terminal

```
pi@raspberrypi:~/Documents $ ls
BlueJ Projects   Greenfoot Projects   Scratch Projects
```

As you can see, a different set of contents has been listed for your new directory. The command `cd ..` takes you back to the folder you just came from. You can think of it as moving you back "up" the directory path. So running `cd ..` in the *Documents* folder would take you back to the home directory, for example.

One final (and important) command that you should know is `shutdown`. The Raspberry Pi has no power switch, and you disconnect power to it by pulling out the power cord. Every time before you do this, you need to safely shut down the operating system. Whenever you want to turn off your Pi, run this command:

```
pi@raspberrypi:~ $ sudo shutdown now
```

After this, wait for a few moments and then feel free to remove the power cord.

And with that, I have introduced you to the mechanics of the two most useful Linux terminal commands: `ls` and `cd`. You'll use these a lot in your Raspberry Pi adventures. We'll use many other terminal commands over the course of this book, too, and I'll introduce them to you as we go along.

Accessing Your Raspberry Pi from Another Computer

Setting up a Raspberry Pi is easy, but having it always connected to a monitor can be a hindrance. You don't want a long wire trailing from your robot as it scurries around the floor, so now I'll show you how to access your Raspberry Pi from another computer so you can send your robot instructions wirelessly.

You can access the terminal of a Raspberry Pi remotely from another computer that's on the same network using *SSH*, an internet protocol that stands for *Secure Shell*. You can then input commands to a command line of that computer and see their responses, exactly like you did in the previous section, and send them to your robot.

First, your Raspberry Pi *must* be connected to your local network, either over Wi-Fi or Ethernet, and the other device *must also* be connected to that same network.

Next, you'll need your Pi's *IP address* in order to connect to it later. This is an *Internet Protocol address*, a numerical label given to every device connected to a network that's unique to that device. Think of it like your house address, but for your computer.

To find out your Raspberry Pi's IP address, simply open a terminal and enter the following command:

```
pi@raspberrypi:~ $ ifconfig
```

This will generate a large amount of output text. It might look confusing, but your IP address is in there.

If you're connected to your network over Wi-Fi, scroll down to the wlan0 entry, and the number after inet is your IP address; see Figure 1-18 for mine. If you're connected over Ethernet, you can find your IP address under the eth0 entry. In any case, just look for the magic sequence following the form inet 192.168.1.221 (or something similar) within the output.

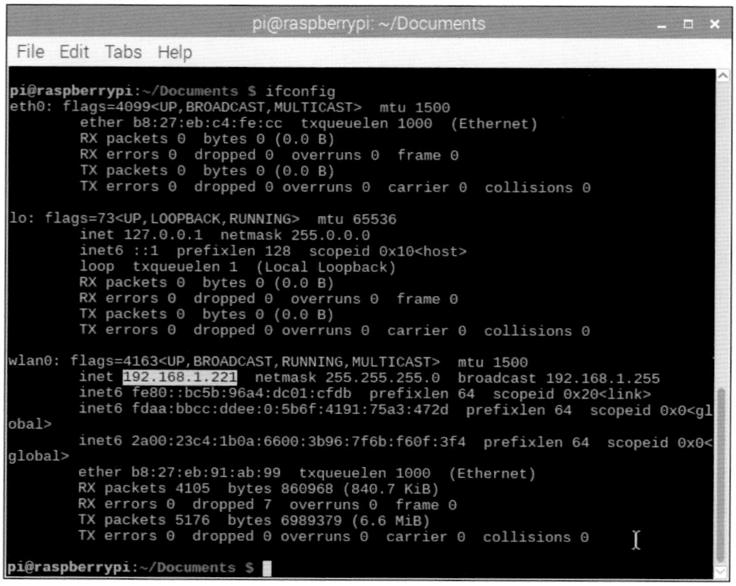

FIGURE 1-18

The output of the ifconfig command. I have highlighted my IP address—this is where you should find yours if you are connected over Wi-Fi.

```
                          pi@raspberrypi: ~/Documents         _  □  ×
  File  Edit  Tabs  Help

pi@raspberrypi:~/Documents $ ifconfig
eth0: flags=4099<UP,BROADCAST,MULTICAST>  mtu 1500
        ether b8:27:eb:c4:fe:cc  txqueuelen 1000  (Ethernet)
        RX packets 0  bytes 0 (0.0 B)
        RX errors 0  dropped 0  overruns 0  frame 0
        TX packets 0  bytes 0 (0.0 B)
        TX errors 0  dropped 0 overruns 0  carrier 0  collisions 0

lo: flags=73<UP,LOOPBACK,RUNNING>  mtu 65536
        inet 127.0.0.1  netmask 255.0.0.0
        inet6 ::1  prefixlen 128  scopeid 0x10<host>
        loop  txqueuelen 1  (Local Loopback)
        RX packets 0  bytes 0 (0.0 B)
        RX errors 0  dropped 0  overruns 0  frame 0
        TX packets 0  bytes 0 (0.0 B)
        TX errors 0  dropped 0 overruns 0  carrier 0  collisions 0

wlan0: flags=4163<UP,BROADCAST,RUNNING,MULTICAST>  mtu 1500
        inet 192.168.1.221  netmask 255.255.255.0  broadcast 192.168.1.255
        inet6 fe80::bc5b:96a4:dc01:cfdb  prefixlen 64  scopeid 0x20<link>
        inet6 fdaa:bbcc:ddee:0:5b6f:4191:75a3:472d  prefixlen 64  scopeid 0x0<gl
obal>
        inet6 2a00:23c4:1b0a:6600:3b96:7f6b:f60f:3f4  prefixlen 64  scopeid 0x0<
global>
        ether b8:27:eb:91:ab:99  txqueuelen 1000  (Ethernet)
        RX packets 4105  bytes 860968 (840.7 KiB)
        RX errors 0  dropped 7  overruns 0  frame 0
        TX packets 5176  bytes 6989379 (6.6 MiB)
        TX errors 0  dropped 0 overruns 0  carrier 0  collisions 0       I

pi@raspberrypi:~/Documents $
```

As you can see, my IP address is 192.168.1.221. Make a note of yours (it will probably be quite similar), as you will use this information to connect to your Pi from your other device.

Next, you need to set up an SSH connection on your other computer. SSH is built into Linux distributions and macOS, but for Windows you'll need some third-party software. I'll cover the process of connecting to your Pi from both Windows and Mac next; you only need to read the section relevant to you and your machine.

Using SSH on Windows

The free software required for Windows is called PuTTY. The installation process is easy and takes only a few minutes:

1.   On your Windows PC, open a web browser and navigate to *http://www.putty.org/*, then follow the link to the download page.

Select the 32-bit or 64-bit option, depending on your machine. This will download the installer.

2. Open the installer and follow the instructions to install the PuTTY software.

3. Once the installer has completed, open the PuTTY application. You'll find a shortcut to PuTTY in the Start menu. Every time you open the application, you'll be presented with a box like the one in Figure 1-19.

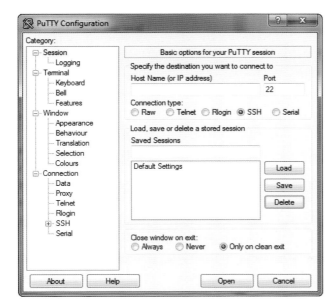

FIGURE 1-19

The PuTTY application and configuration dialog box

4. To connect to your Raspberry Pi, enter your Pi's IP address into the box labeled "Host Name (or IP address)." Below that, ensure that SSH is selected and, unless you know otherwise, keep the Port option set at 22. Click **Open** to start the connection.

5. If this is your first time connecting to a new Pi, you'll probably see a security warning; simply click **Yes** and agree. After this, a password prompt will appear. Type in your Pi's password and press ENTER; if you didn't change it earlier, your password will be the default *raspberry*.

You now have a full Raspberry Pi terminal available for you to use, as shown in Figure 1-20. Why not try out some of the commands you learned in the previous section?

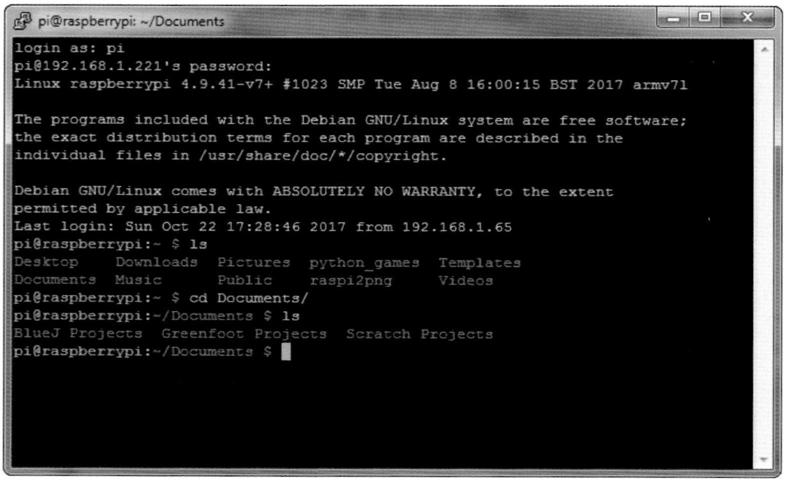

FIGURE 1-20

The PuTTY shell

```
pi@raspberrypi: ~/Documents

login as: pi
pi@192.168.1.221's password:
Linux raspberrypi 4.9.41-v7+ #1023 SMP Tue Aug 8 16:00:15 BST 2017 armv7l

The programs included with the Debian GNU/Linux system are free software;
the exact distribution terms for each program are described in the
individual files in /usr/share/doc/*/copyright.

Debian GNU/Linux comes with ABSOLUTELY NO WARRANTY, to the extent
permitted by applicable law.
Last login: Sun Oct 22 17:28:46 2017 from 192.168.1.65
pi@raspberrypi:~ $ ls
Desktop     Downloads   Pictures   python_games   Templates
Documents   Music       Public     raspi2png      Videos
pi@raspberrypi:~ $ cd Documents/
pi@raspberrypi:~/Documents $ ls
BlueJ Projects   Greenfoot Projects   Scratch Projects
pi@raspberrypi:~/Documents $ ▮
```

Using SSH on macOS

No extra software is required to use SSH on any Mac computer. Instead, you can use the Mac's very own Terminal program!

1. From anywhere on your Mac, hold the COMMAND key and space-bar to open up a Spotlight Search. Type in **terminal** and press ENTER. A command line box should open up.

2. To connect to your Raspberry Pi, enter the following command into Terminal, replacing the text ***your_ip_address*** with your actual IP address you found earlier:

    ```
    $ ssh pi@your_ip_address
    ```

    As you can see in Figure 1-21, the command I entered was:

    ```
    $ ssh pi@192.168.1.221
    ```

3. If this is the first time you've connected to a new Pi, you'll probably see a security warning; simply enter **yes** to agree to it. You will then have to enter your Raspberry Pi's password and press ENTER; if you didn't change it earlier, then your password will be the default *raspberry*.

FIGURE 1-21

The Mac Terminal SSH connection process

You should now have a full Raspberry Pi terminal available for you to use! Try out some of the commands you learned in the previous section!

WHAT IS PROGRAMMING?

You're almost ready to move on and start building robots! The last thing I want to do in this chapter is introduce you to a few programming concepts.

Programming is the process of writing a set of instructions that tell a computer how to perform a task. You can program video games, apps, or even, in our case, robots. The only thing that limits what you can do with computers and programming is your imagination!

Over the course of this book, I'll show you how to wire up electronic components and then program them to do exactly what you want. If you have never done any programming before, don't worry. I will explain everything as we progress through the book. You can also download all the code you'll need from *https://nostarch.com/raspirobots/* if you ever get stuck.

Introducing Python

There are many different programming languages, each with its pros and cons. The language we'll use throughout this book is called *Python*. Python is a *high-level* programming language, which means that it's more similar to human language than many other programming languages, so it's both easy to understand and powerful.

You'll be writing Python programs and then executing them from the terminal. This means that you'll be able to use SSH to program your Raspberry Pi from another computer without ever needing to plug it in to a monitor.

Writing Your First Python Program

For your first program, it is geek tradition to create a computer program that *outputs*, or displays, "Hello, world!" to the user. This is a simple task that will help you get the hang of how to write and run Python programs.

1. From the terminal, enter the following line:

```
pi@raspberrypi:~ $ mkdir robot
```

2. This command will create a new directory called *robot* for you to store your own Python programs. We'll store all our programs in this directory from now on. Change into your new directory with this command:

```
pi@raspberrypi:~ $ cd robot
```

3. Next, from your new *robot* directory, enter this line:

```
pi@raspberrypi:~/robot $ nano hello_world.py
```

4. Nano is a *text editor*. Text editors are simple software that allow you to make, review, and change files with text in them. Nano lets you create and edit programs directly in the terminal, without opening any other windows. This command uses Nano to create a new file called *hello_world.py*—the file extension *.py* tells Nano that this is a Python program. From this command an empty *hello_world.py* file has been created and saved in our *robot* directory. This file is now open in your terminal.

5. With Nano open in your terminal, you enter instructions to save into *hello_world.py*. Enter the following single line of Python code:

```
print("Hello, world!")
```

6. The command `print()` is a Python function. In programming, a *function* is an instruction that performs a task. All `print()` does is output to the terminal whatever text you give it! In our case we have given it the text "`Hello, world!`". In programming, any series of characters between two quote marks is known as a *string*.

7. With your one-line program now complete, you need to exit and save your work. To do this in Nano, press CTRL-X. You will then be asked whether you want to save the changes you have made. Press Y to say yes. Nano will then ask you what filename you want to save these instructions to, which in our case is *hello_world.py*. Press ENTER to confirm this. You should now be back at the command line.

8. You've created your program, so now you need to run it. To run a Python program from the terminal, simply enter **python3** followed by the name of the file to run. Enter the following to execute your program:

```
pi@raspberrypi:~/robot $ python3 hello_world.py
```

You'll see the string "`Hello, world!`" displayed in your terminal window, as in Figure 1-22.

FIGURE 1-22

The output of your first Python program

You may have noticed the 3 in the command `python3 helloworld.py`. This is telling your Pi to execute the file using Python 3, rather than Python 2. Python 3 is the newest version of Python, and while version 2 is still used a lot, using Python 3 is the preferred option. There's not a lot of difference, but there are some syntax and feature differences between them. All of the projects in this book will use Python 3.

SUMMARY

We've covered a lot in this chapter! You've gotten to know your new Raspberry Pi, set it up, and had your first taste of both the terminal and Python programming. You're now able to access and use your Pi remotely and understand how to do things like change preferences and set up an SD card.

In the next chapter I'll cover the basics of electronics and electricity, and you'll start doing some simple building in the form of mini-projects, like flashing LEDs, and more. This will give you the foundation of knowledge that you'll need before we move on to making robots!

2
ELECTRONICS BASICS

ELECTRONICS IS THE SCIENCE OF CONTROLLING AND MANIPULATING ELECTRICAL ENERGY TO DO SOMETHING USEFUL. IT'S ABOUT MAKING ELECTRONIC COMPONENTS LIKE LIGHTS, SENSORS, AND MOTORS DO EXACTLY WHAT YOU WANT THEM TO DO.

Many innovations stem from the different fields of electronics. Most interesting for us is the branch of *robotics*. To make your own robots, you'll need to understand the basics of electronics and bend this knowledge to your will! In this chapter, I'll give you your first taste of electronics in the form of two projects. You'll program an LED (light-emitting diode) to blink at regular intervals, and then wire up a button to print a message to your terminal when it's pressed. You'll be blinking LEDs and controlling the physical world in no time at all!

WHAT IS ELECTRICITY?

Electricity is everywhere in our day-to-day lives: electric currents are used to power electrical components and appliances, like the lights in your house, your TV screen, your toaster, and the motors of a Raspberry Pi robot. But what actually *is* electricity?

Electricity starts with *atoms*. Everything in the world is made out of billions of tiny atoms—even you! And as you may have learned in science class, atoms themselves are composed of three particles: *protons*, *neutrons*, and *electrons*. The protons and neutrons sit together in the center of the atom to form the atom's *nucleus*, and the electrons orbit that nucleus, as shown in Figure 2-1.

FIGURE 2-1

A diagram of an atom

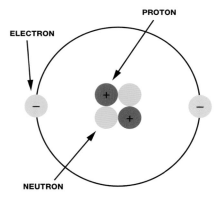

Protons and electrons each have *electric charge*, which is a fundamental property of matter. Protons are positively charged, and electrons are negatively charged. Neutrons have no charge; that is, they are neutral. You may have heard the saying "opposites attract," and that applies here. Because protons and electrons have opposite charges, they are attracted to each other and stay together, forming the atoms that make up everything around you.

Atoms come in many different arrangements called *elements*. Each element is defined by the number of protons, electrons, and neutrons each atom contains. For example, the element copper usually has 29 protons and 35 neutrons, while gold has 79 protons and 118 neutrons. All metals, like copper, gold, and iron, are made out of collections of atoms all pressed up against each other. Some of these materials are *conductive*, which means that, when given energy, the electrons from one atom can move to the next atom. This causes a *flow of charge* in the material, known as an electric *current*. The number of electrons flowing through a point in a material at any given second is the size of the electric current, which is measured in *amperes (A)*.

For an electric current to flow, there must be a complete *circuit*. A circuit is a closed path, like a loop, around which an electric current moves. The circuit must be made of conductive material for the electricity to move through, and any gap in the circuit means the electricity cannot flow.

The circuit needs a source of energy to "push" the electric current around. This can be a battery, a solar panel, electrical mains, or any number of things. Crucially, these sources provide a *potential difference*, known as a *voltage*. A voltage simply pushes electrons through a conductor, such as copper wire, and the strength of a voltage is measured in *volts (V)*.

Power sources have a positive and negative terminal. In a simple circuit, like the one shown in Figure 2-2, the terminals of a battery could be connected by a thick copper wire. Electrons are negatively charged and are therefore attracted to the positive terminal of the battery, so they travel through the circuit from the negative end to the positive end, pushed along by the voltage.

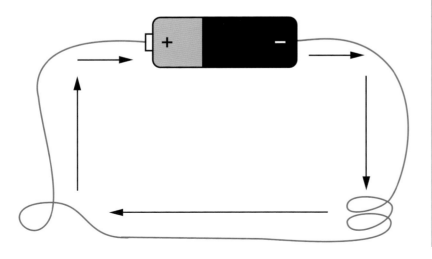

FIGURE 2-2

A circuit showing the flow of charge around a thick wire connected to the positive and negative terminals of a battery

Although the electrons flow from *negative to positive*, it is convention to think of the current flowing from *positive to negative*. The battery in this circuit has a fixed voltage. If this voltage is increased, more electrons would be pushed around the circuit and the current would be larger. Conversely, if this voltage is decreased, fewer electrons would be pushed around the circuit and the current would be smaller.

Resistance

Now that you have an understanding of circuits, we need to add another ingredient into the mix: *resistance*. Resistance simply reduces current. Outside the laboratory every material has some amount of resistance, which is measured in *ohms (Ω)*. One way to think about resistance is to imagine a water pipe. The water flowing through the pipe is like electric current flowing through a copper wire. Imagine the water pipe has one end higher than the other. The water at the higher end of the pipe has more energy (potential energy) than water at the lower end. If the pipe is level, no water will flow. If the pipe is slightly sloping, a small flow will occur. The actual amount that flows depends on both the difference in height of the ends above ground and how wide the pipe is. The height difference of the pipe is like potential difference, or voltage.

Resistance, on the other hand, is like something squeezing the pipe and affecting how wide it is: the more it is squeezed, the less water is able to flow through it (see Figure 2-3). This translates to less electric current flowing through the circuit.

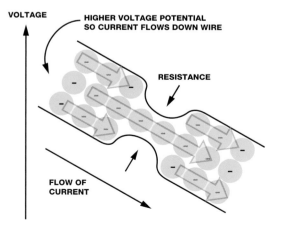

Therefore, three ingredients make up an electric circuit: voltage, current, and resistance. They all seem to be pretty closely connected, right? You may even think that there must be a certain *mathematical* connection or *law* relating to them—and you'd be right.

Ohm's Law of Electricity

Ohm's law deals with the relationship between voltage, current, and resistance. It states that the *voltage across a conductor is proportional to the current running through it*.

Let's break this down to see what it means. In a circuit, voltage is simply equal to current multiplied by resistance. We use *V* to stand for voltage, *I* for current, and *R* for resistance. So, the equation for voltage is written as follows:

$$V = I \times R$$

As with any mathematical equation, you can rearrange it to work out the equations for the other terms. For example, from Ohm's law we know that the current in a circuit is equal to the voltage divided by the resistance. When you rearrange the equation for current and resistance, you get the following equations:

$$I = \frac{V}{R}$$

$$R = \frac{V}{I}$$

If all of this is a little confusing, don't worry! As you make your own circuits, it will become easier to understand. Now that we have covered some of the basics of electricity and electronics, let's get making!

MAKING AN LED BLINK: RASPBERRY PI GPIO OUTPUT

Just as "Hello, world!" is a traditional first program, making an LED blink is a traditional first electronics project since it very neatly demonstrates using the GPIO pins as outputs. This project will be your introduction to using your Pi's GPIO pins. Before we begin, you might have some questions.

First, what is an LED? Short for *light-emitting diode*, an LED is a component that gives off light when an electric current is passed through it. LEDs are the modern equivalent of an old light bulb, but they use less power, don't get hot, and have a longer life.

The Parts List

For your first foray into electronics, you're going to need a few extra things besides the Raspberry Pi you set up previously. Here's what you'll need for this project:

NOTE

For guidance about where to buy and source these parts, check the Introduction.

- A breadboard
- An LED (color of your choice)
- An appropriate resistor
- Jumper wires

Before we wire these components up, I'll explain a little more about how they work and why you need them.

Breadboard

An electronics *breadboard* allows you to connect electronic components without having to fuse them together permanently (something that is called *soldering*; see "How to Solder" on page 204). This means you can quickly prototype circuits by inserting components into a breadboard's holes. The space between the holes of a breadboard is standardized (2.54 mm/0.1 inches), so all breadboard-friendly components should fit with no trouble. Breadboards come in several sizes with different numbers of holes (also known as *points*). I would recommend a 400-point breadboard, like the one in Figure 2-4.

FIGURE 2-4

A 400-point breadboard and a diagram of how the rows and columns are connected to each other

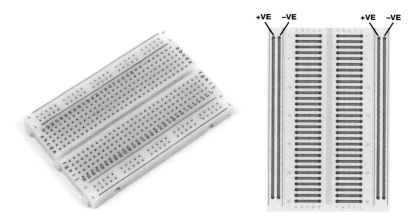

You can see in Figure 2-4 how the rows and columns of the breadboard are internally connected with metal strips. So, if you put one component into a row and put something else into the same row, for example, they will be connected in a circuit.

LEDs

LEDs come in all different shapes, sizes, and colors. Fortunately, they are also incredibly cheap. When bought in bulk they are quite literally *pennies* each. Make sure that your LED has two legs that can be arranged to fit in your breadboard, as shown in Figure 2-5.

FIGURE 2-5
A blue LED

Feel free to buy an LED in any color you wish—I have gone for blue. Make sure to check the voltage specification for the LED you buy. You need to make sure that the voltage required to light up the LED is less than 3.3 V. This is often referred to as the *forward voltage*. You can usually find this information in the online listing for your LED. The forward voltage of my LED is 2.5 V. The Raspberry Pi's GPIO pins work at 3.3 V, so if your LED has a forward voltage of 5 V, for example, your Pi won't be able to light it up!

You also need to find out the forward current of your LED. The *forward current* is the recommended current to run through your component. My LED has a recommended forward current of 30 mA (*milliamps* are one thousandth of an amp), which is the equivalent of 0.03 A. If you provide less current than recommended, your LED won't be very bright; if you provide too much current, it might blow up (you'll hear a small pop when this happens). This information will also most likely be in the LED's internet listing or packaging. If you aren't sure about the specifics of your LED, don't worry—small, cheap ones are usually just fine for our use. If you have no information about your LED, just assume that the forward voltage is around 2 V and the forward current is about 20 mA.

Resistors

To avoid overloading our LEDs, we'll use a *resistor*. Every material has resistance, but resistor components are designed specifically to create pure resistance in circuits.

LEDs, and most components, are quite sensitive to the amount of current that flows through them. If you were to connect an LED directly to a battery and create a circuit without a resistor, the amount of current that would flow through the LED could be large enough to cause it to overheat. A resistor lowers the current through the LED to prevent this from happening.

Resistors come in different values denoted by colored bands, which you can see in Figure 2-6. Take a look at the resistor guide on page 202 to learn what these bands mean and how to read them.

page 202

FIGURE 2-6

A resistor

To find out what resistor you need, you'll have to apply Ohm's law! From the equation you saw earlier, you know that resistance is equal to the voltage divided by the current, or $R = V/I$. In our case, the voltage is the difference between the voltage the Pi supplies, 3.3 V, and the forward voltage of the LED: it is the total source volts *minus* the LED volts. For me, that is 3.3 V – 3 V = 0.3 V. You should use *your* forward voltage here instead, or 2 V if you don't know it.

The current is the forward current of your LED. For me that is 0.03 A. Make sure that this value is in amps, not milliamps!

I can work out the value of the resistor I need to lower the current to 0.03 A by simply calculating the following equation: 0.3 / 0.03 = 10. This means that I will need a resistor of approximately 10 Ω. Often you won't be able to find a resistor value for the specific number you've calculated. That's okay: in most cases, you can simply use the nearest valued resistor you can find. For my LED I was lucky and had a resistor that matched the value I needed exactly. I am using the 10 Ω resistor pictured in Figure 2-6.

If you're still unsure about the forward voltage and forward current of your LED, just err on the side of caution and fit a sensibly

NOTE

I recommend buying a selection of resistors, which are normally organized into something that looks like a book. That way, you'll have a resistor for every occasion and won't have to buy them individually.

large resistor of at least 100 Ω into your circuit. If the LED is too dim, downsize the resistor until you get to a suitable level of brightness (dim enough to not hurt your eyes is a good rule of thumb). Don't try to do this the other way around: you can't unexplode an LED!

Jumper Wires

Finally, you'll need some wires to connect everything up. Specifically, you'll need *jumper wires*, which are breadboard-friendly wires that allow you to connect things to the Pi's GPIO pins. You can see some examples of jumper wires in Figure 2-7.

FIGURE 2-7
A collection of jumper wires

The ends of jumper wires are either *male* or *female*. A male end (often abbreviated as M) has a wire sticking out of it that you can insert into a breadboard's holes. A female end (abbreviated as F) instead has a hole into which you place a wire. I would recommend buying a variety so that you have a jumper wire for all situations. We'll be using a lot of these throughout the book! In Figure 2-7, you can see my collection of M-M, M-F, and F-F jumper wires. For making an LED blink, we'll need two M-F jumper wires.

Wiring Up Your LED

Now that you've collected your parts, it's time to wire up your LED and create your first circuit! You'll wire up your circuit as shown in Figure 2-8, so you can use this diagram as a reference as you go through the instructions.

FIGURE 2-8

Breadboard diagram for
wiring up an LED

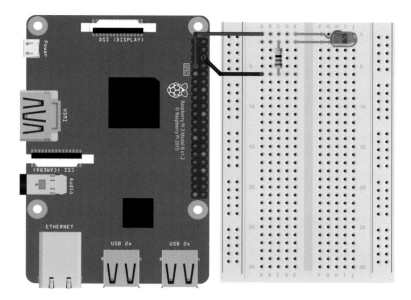

Depending on the breadboard you have, your circuit may look slightly different. To make sure your connections are correct, follow these instructions:

1. Insert the LED into the breadboard so that each leg is in a different row. If you put the LED's legs into the same row, they'll be connected to each other, but won't be connected to anything else. LEDs have a positive and negative side, which you need to align with the flow of the current. The long leg of the LED is the positive side—called the *anode*. The short leg is the negative side—the *cathode*. The LED bulb will usually be flat on the side of the cathode as an extra indicator.

2. Insert one leg of your resistor into the same row on the breadboard as your LED's shorter leg. Putting the resistor leg in the same row as your LED connects the two in a circuit. Connect the other leg of the resistor to any of the other points of the board.

3. Now, with your Raspberry Pi turned off, insert the male end of one of your M-F jumper wires into the breadboard, in the same row as the long leg of your LED. Locate physical pin 7 on your Raspberry Pi, also known as the *GPIO/BCM 4* pin (see "Raspberry Pi GPIO Diagram" on page 200 for an image of this), and connect the female end of the wire to it.

4. Finally, insert the male end of your other M-F jumper wire into the row of the breadboard that contains only one leg of the resistor and none of the LED's legs. Then connect the female end to physical pin 6 on your Raspberry Pi. This is one of the *ground* pins. You can think of ground as the negative terminal of a battery. It is just the lower side of a voltage.

Programming Your Raspberry Pi to Blink Your LED

You should now have your circuit wired up, so boot up your Raspberry Pi and log in. It's time to write a program to blink that LED!

From the terminal, navigate from the home directory into the folder you created in Chapter 1 with the command:

```
pi@raspberrypi:~ $ cd robot
```

Now you'll create a new file and write a Python program to control your LED. Pick whatever name you like for your file, but ensure your filename ends with *.py*. I've called mine *blink.py*. The following command creates a new file and opens the Nano text editor:

```
pi@raspberrypi:~/robot $ nano blink.py
```

You'll now find yourself in a Nano text editor identical to the one you came across in Chapter 1.

Enter the code in Listing 2-1 to instruct your LED to flash on and off (the numbers in circles don't actually appear in the program, but we'll be using them for reference).

NOTE

When you power on your Raspberry Pi, your LED may be off, on, or even dimly lit. Don't worry! Your LED is fine in any of these states. You haven't yet instructed the pin to be a certain state, so your pin isn't quite sure what to do yet.

LISTING 2-1

Program to blink
an LED

```
❶ import gpiozero
   import time

❷ led = gpiozero.LED(4)

❸ while True:
❹     led.on()
❺     time.sleep(1)
❻     led.off()
❼     time.sleep(1)
```

This eight-line Python program is easy to understand when you look at it one line at a time, so let's break it down.

Python is an *interpreted* programming language, meaning when this code is run, your Raspberry Pi (or any other computer) will execute your program line by line, starting at the top and moving down in a logical manner. That means the order of your code matters.

Python comes with all sorts of built-in abilities. For example, in Chapter 1 you *printed* text to the terminal, a capability Python has by default. There are hundreds of other things Python can do, but some abilities need to be imported from external sources. For example, Python is not able to control your Pi's GPIO pins on its own, so we import a library called GPIO Zero ❶. In programming, a *library* is a collection of functions a program can use. By importing a library, we bring these functions into the current program for our own use. The GPIO Zero library was created by the Raspberry Pi Foundation to give programmers a simple GPIO interface in Python. Importing this library enables your program to control your Pi's GPIO pins! Note that it's actually called gpiozero in the programming language, though, as we can't include spaces in library names and the convention is to use lowercase.

On the next line we import the time library, which allows your Python program to control timings. For example, you'll be able to pause the code, which will be very useful in our case!

Next, we make a variable ❷. In programming, *variables* are names used to store information to be referenced and manipulated in a program. They provide a way of labeling data, and they make code simpler, easier to understand, and more efficient.

In this case, we've created a variable called led that references the LED software from the GPIO Zero library. We give the LED() function the value 4 in parentheses to show that we are referring to an LED on GPIO/BCM pin 4. When we call led later in the program, the Pi knows we mean this pin.

Then we begin a `while` loop ❸, which is a *conditional statement* that will keep running the code inside it until the condition is no longer met. In simple English, we're telling the loop: while this condition is true, keep running the code. In this case, the condition is simply `True`. The `True` condition will always be true and will never be false, so the `while` loop will go around and around indefinitely. This is useful to us, as we'll be able to write the code to make the LED flash once, and the loop will take care of making the LED flash over and over again.

Within the `while` loop, you also come across a key structural feature of Python: *indentation*. Python knows that all the code indented by the same number of spaces belongs to the same group of code, known as a *block*. The four lines following the `while` loop are indented four spaces each; as long as the condition is true, the loop will run that whole block of code.

You can create indentation in different ways. Some people use two spaces, four spaces, or a TAB. You can use any method you like as long as you stay consistent throughout your Python program. I'm a TAB person myself.

At ❹, you switch the LED on using the command `led.on()`. Remember that `led` refers to the pin we connected the LED to and now we're telling that pin to be "on." The dot (`.`) separates the thing we're talking about, in this case the LED, from what we're asking it to do, in this case be turned on. Turning on a GPIO pin is also known as bringing that pin *high*, since the Raspberry Pi will apply a voltage of 3.3 V across your circuit when this line of code runs.

Next we use a `sleep()` statement ❺ to tell your program to pause for whatever number of seconds you give to it in parentheses. In this case, we entered a value of 1, so the program sleeps for just 1 second. After this, you switch the LED off using the command `led.off()` ❻. Repeat the `sleep()` statement at ❼ to make the program wait for another second before looping back around again to the start of the `while` loop. This sequence of on-wait-off-wait continues indefinitely.

Once you've finished entering the code for your program, you can exit the Nano text editor and save your work. To do this, press CTRL-X. You will then be asked whether you would like to save the changes you have made. Press the Y key to say yes. Nano will then prompt you for the filename you would like to write to, which in our case should be *blink.py* or the filename you entered when you opened the Nano editor. Press ENTER to confirm the filename.

Running Your Program: Make Your LED Blink

Now that you understand how your program works, it's time to run it. You'll follow the same process to execute your program as you did for the *helloworld.py* program you created in Chapter 1. Enter the following code into the Raspberry Pi's prompt:

```
pi@raspberrypi:~/robot $ python3 blink.py
```

Your LED should now start to blink on and off at regular intervals (see Figure 2-9). Congratulations, you've just successfully interfaced your Raspberry Pi with the outside world!

FIGURE 2-9

A happily blinking LED connected to the Raspberry Pi

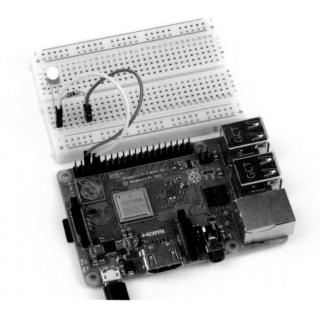

To kill your program and stop the blinking LED, press CTRL-C.

TROUBLESHOOTING GUIDE: WHY ISN'T THE LED BLINKING?

If your LED isn't blinking, don't panic. You can likely fix your circuit with a little troubleshooting. First, check whether your LED is inserted with the legs the right way around. Because LEDs are a form of *diode*, current flows through them only in one direction, so if you have your LED in the wrong way it won't light up! Instead, it'll just do nothing. Go back to the instructions and make sure you followed them accurately.

If this doesn't fix your problem, check the rest of your circuit. Is everything connected properly? Are all of the wires firmly in place? Check that you have wired up your circuit to the correct pins of your Pi's GPIO port—this is an easy mistake to make!

If you're convinced that your circuit is sound and your LED and resistor are appropriate (as explained in the parts list), then you may have a software issue. When you ran the program, did it crash? Did you get an error message? You may have made an error when copying the code from this book. Go back and check, or grab the code files from *https://nostarch.com/raspirobots/* and run the *blink.py* file from there instead.

The GPIO Zero library is included by default in all new Raspbian releases, but if you are running an older version of Raspbian, you may need to install the library manually. To do this, enter the command:

```
pi@raspberrypi:~/robot $ sudo apt-get update
```

followed by the command:

```
pi@raspberrypi:~/robot $ sudo apt-get install python3-gpiozero
python-gpiozero
```

Challenge Yourself: Change the Timing

Take a look at the code you used to make your LED blink. What would happen if you modified some of it? For example, you could experiment by changing the timing of the `sleep()` statements and seeing what different patterns you can make! Play around a bit to see what effects your changes have.

INPUT FROM A BUTTON: RASPBERRY PI GPIO INPUT

Blinking an LED is the perfect first experiment for the world of electronics and physical computing with your Raspberry Pi, but it demonstrates only the *output* aspect of what the Pi's GPIO pins can do. GPIO pins can also take *input*, meaning they can take data from

NOTE

*If you want to shut down your Raspberry Pi, you should do so safely in software before yanking out the power cord. To commence a power down sequence, use the command **sudo shutdown now**. Wait a few seconds before pulling out the power cord. Or, choose the shutdown option in the main menu from the GUI if you're using a directly connected screen.*

the outside world and react to it. In this section you'll wire a button up to your Raspberry Pi and write a program that is triggered whenever that button is pressed.

Explaining the Parts List

The projects in this book build upon each other, so in addition to your Raspberry Pi and breadboard, for this project you'll need a button and two M-F jumper wires.

Buttons come in hundreds of different shapes, sizes, and varieties. For this project, you'll need to acquire a breadboard-friendly *momentary push button*, like the one shown in Figure 2-10.

FIGURE 2-10

A four-legged momentary push button and a diagram of the leg pairs

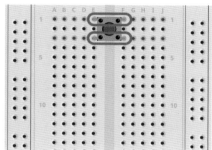

Buttons are often referred to as *switches*, and most have either two or four legs that connect to the points of a breadboard. The function of a momentary switch is simple: when the button is pressed, the contacts inside the button join together and complete the circuit. When the button is released, the contacts inside are separated, so the circuit is incomplete and no current flows. That means your circuit is connected only while the button is pressed; that's why it's called "momentary"!

For buttons with two legs, it's obvious that when the button is pressed both sides are connected. For buttons with four legs, it's a little bit more complicated. The legs are set up in pairs, so you only actually need to worry about two of them. Usually, opposite legs, like the ones indicated in Figure 2-10, are coupled, which means only one leg of each pair needs to be connected to the circuit. If you are unsure of which legs are pairs, check your button's specification.

Wiring Up Your Button

With your parts ready, you can now wire up your button. Use the breadboard diagram in Figure 2-11 as a reference.

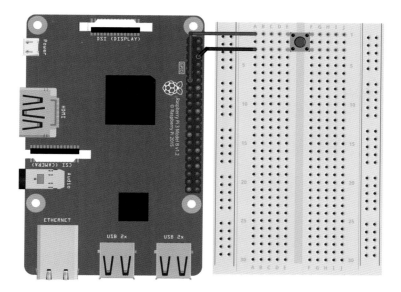

FIGURE 2-11

Breadboard diagram for wiring up a button

If you've just finished the previous project and still have an LED wired up to your Pi, feel free to take the circuit apart or wire up your button on a different part of your breadboard—just make sure that you use different rows of the breadboard for each project. Once you're ready to set up your next circuit, follow these instructions:

1. Insert your push button into your breadboard and ensure that each leg is in its own row. You'll need to insert the button with two legs on either side of the row divider in the middle of your breadboard to accomplish this.

2. With your Raspberry Pi turned off, use a jumper wire to connect one leg on one side of the divider to the ground pin on your Raspberry Pi (physical pin 6).

3. Use another jumper wire to connect the other leg on the same side of the divider to pin 11 (BCM pin 17) on your Raspberry Pi. Check the pinout on page 201 if you're not sure which pin is which.

With your Pi setup like mine in Figure 2-12, you now need to write a program to take input from the button when it is pressed.

Programming Your Raspberry Pi to Display Input from Your Button

From the terminal, make sure you're in the folder you created in
Chapter 1. If you just completed the previous mini-project you most
likely already are, but if not, just navigate from the home directory into
the *robot* directory with this command:

```
pi@raspberrypi:~ $ cd robot
```

Now create a new Python program to get input from your button
in Nano. I've created one and called it *button.py* as follows:

```
pi@raspberrypi:~/robot $ nano button.py
```

You'll find yourself in a familiar blank Nano interface. Enter the
code in Listing 2-2 to get your button working.

```
❶ import gpiozero

❷ button = gpiozero.Button(17)

  while True:
  ❸   if button.is_pressed:
  ❹       print("Button is pressed!")
  ❺   else:
          print("Button is not pressed!")
```

We import the GPIO Zero library in exactly the same way as we did for the blinking LED project ❶. All of the electronics projects in this book will use this library, so you can expect to see this line in all the programs in this book!

We then create a variable called button ❷, assigning some button software from the GPIO Zero library to that variable and making sure to specify in the parentheses that the button is connected to BCM pin 17.

We then start a while loop as we did in our LED program, but unlike our previous program, the first line in this block starts an if/else statement. In programming, if statements are conditional statements that activate some code when a condition is satisfied. Our if statement almost directly translates into English as "if the button is pressed, do the following" ❸. While the if statement is true, the indented line ❹ is executed and a string of text is printed to the terminal, telling us the button has been pressed.

Usually, but not always, when there is a conditional if statement, there is often an else statement, because if the button isn't pressed, something *else* must happen! The else statement ❺ can be translated into English as "if anything else, do the following." The indented block of code following the else statement is then executed and prints to the terminal that the button hasn't been pressed.

Because the if/else statements are in a while loop whose condition is True, the code continues to run forever unless the program is stopped. Once you have finished entering your program, you can exit the Nano text editor as usual: press CTRL+X, save the changes you have made by pressing the Y key at the prompt and then press ENTER to confirm the filename as *button.py*.

Running Your Program: Get Input from Your Button

To run your program, simply enter the following into the terminal:

```
pi@raspberrypi:~/robot $ python3 button.py
```

You should now see the statement Button is not pressed! repeatedly in the terminal. When you press your button, the statement Button is pressed! should be printed to your terminal.

```
pi@raspberrypi:~/robot $ python3 button.py
Button is not pressed!
Button is not pressed!
Button is pressed!
```

TROUBLESHOOTING GUIDE:
WHY ISN'T THE BUTTON WORKING?

Just as in the previous mini-project, if your button isn't working, don't worry! First, check your circuit. Is everything wired up correctly? Are the wires firmly in the breadboard? Have you connected your button in the right way? If your switch is slightly different from my model, have you looked into the details of how it is different? If you think that your circuit is okay, you may have made an error when copying the code from this book. Go back and check, or grab the code files from *https://nostarch.com/raspirobots/*.

To kill your program, press CTRL-C.

Challenge Yourself: Combine Your Button and LED Programs

See if you can combine both of the mini-projects in this chapter. Try to create a program that flashes an LED when the button is held down, or a program that keeps the LED turned on until the button is pressed.

SUMMARY

In this chapter you've had your first taste of electronics in the form of two mini-projects. You've learned a lot of theory and fundamentals, all of which are going to come in handy as you build your robot over the next chapters. I've also introduced some key programming techniques and concepts—we'll be using plenty of if/else statements and loops in the future.

In the next chapter I'll guide you through the process of making your first robot. We'll cover the materials and tools you'll need as well as the construction process itself.

3
BUILDING YOUR ROBOT

ROBOTS COME IN EVERY SHAPE, SIZE, AND DESIGN YOU COULD POSSIBLY IMAGINE— FROM THE ROBOTIC ARM ON THE INTERNATIONAL SPACE STATION TO TOYS BUILT FOR ENTERTAINMENT.

Robots can be customized for specific tasks, but in order to do this, you'll first need to understand and build a robot's basic components. In this chapter, I'll show you how to make the base robot that you'll modify and improve for the rest of the book. Once we've programmed it, this robot will be able to move around according to your instructions. In later chapters, we'll add sensors, lights, and a camera to make your robot flashier and smarter!

YOUR FIRST ROBOT

Robots that move around can be split into two distinct groups: ones with wheels and ones without. Those without wheels are often *humanoid*, meaning they have two legs and resemble a human being, or might be based on animals, like dogs (see some examples in Figure 3-1). These robots are usually *very* difficult both to create and to program because the builders have to take into account balance, movement, and a huge range of other factors.

FIGURE 3-1
Robots from Boston
Dynamics

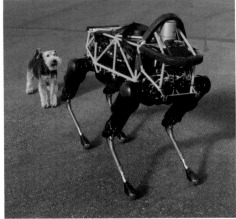

In contrast, robots that use wheels (or tracks) make up the vast chunk of real-world robots and don't have complicated balancing issues, which makes them perfect for hobbyists and makers like us. One famous wheeled robot is NASA's *Curiosity* rover, which has six wheels and has been moving around Mars since 2012, discovering and doing amazing science in the process!

For your first robot, you'll create a two-wheeled robot like the one shown in Figure 3-2. Two-wheeled robots are a great starting point in the world of robotics. Your robot will be able to go forward and backward, and turn left and right using one motor for each wheel.

The two-wheeled robot you'll make in this chapter has just the right mix of features—it's affordable, accessible, and maneuverable!

WHAT YOU'LL NEED

In addition to your Raspberry Pi, you'll need some other basic elements to assemble your robot. As with all things in the maker-world, you have a *huge* number of choices for the materials and components you can use based on what you have available.

If this is your first time working with the Raspberry Pi and electronics, I'd suggest getting the same parts I have, as listed here, so you can follow the instructions word for word. If you have some experience and your own ideas, then feel free to explore; you don't have to stick to what I recommend.

Breadboard I recommend a 400-point breadboard with power rails like the one used in Chapter 2.

Jumper wires I recommend a variety pack of breadboard wires with different colors and lengths.

Chassis This is the robot's body, and should be at least 6 inches by 5.5 inches. I'm using LEGO (see the next section for options).

Two brushed DC motors Use 5 V to 9 V, 100 mA to 500 mA motors with tires and an integrated gearbox.

Battery holder Find one that fits six AA batteries.

Six AA batteries   Either disposable or rechargeable is fine; I recommend Panasonic's Eneloop rechargeable batteries.

LM2596 buck converter   This is a step-down voltage buck converter.

L293D motor driver   This is a motor controller integrated chip.

You will also need a variety of screwdrivers, a hot glue gun, a multimeter, and a soldering iron.

The next few sections provide more details on each component and what it does. If you want to skip straight to building your robot, go to "Assembling Your Robot" on page 60.

Chassis

The *chassis* is the base frame of the robot; you can think of it as the robot's body. The chassis forms the platform on which you'll mount your Raspberry Pi and other parts.

You can create your chassis with a number of different materials, but any design you make will need to satisfy the following three criteria. The chassis needs to be:

A strong, stable platform   All of your robot's electronics will be mounted to the chassis, so you need to make sure it isn't fragile.

At least 6 inches by 5.5 inches   As you progress through this book, you're going to add more and more components to your build, so you'll need space for growth. Don't worry about making your robot chassis *massive*, but make sure it isn't tiny, either!

Easy to modify   Having a chassis that is easy to modify, expand, and change means that you can customize your robot even more in the future!

LEGO is my preferred material for first-time robot building. Everybody's favorite toy bricks can make the perfect robot chassis. With an immense and accessible array of bricks and pieces, you can easily make chassis designs of all shapes and sizes with LEGO parts. In this chapter I'll show you how I have constructed my chassis using the parts in Figure 3-3.

Cardboard is another good option because you can get it anywhere and it's easy to work with. Some simple cutting, folding, and gluing can shape your cardboard into all sorts of chassis shapes. The thicker the card, the better! You can make a solid card-based chassis from any recycled cardboard box—a shoebox, Amazon packaging, a cereal box, you name it!

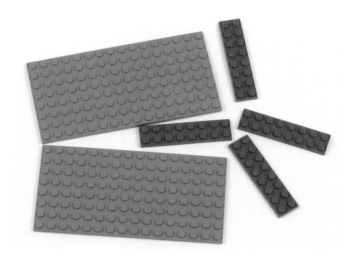

FIGURE 3-3
The LEGO pieces I used to make the chassis of my robot from Figure 3-2

If you have access to some simple wood-crafting tools, like a wood saw, then making a robot chassis out of wood may be a good idea. Inexpensive wood and wood composites, like like pine and medium-density fiberboard (MDF), make a strong, stable chassis at a low price.

A custom plastic chassis can form the basis of a sweet-looking robot. You can purchase acrylic (Plexiglas, Lucite, Perspex) sheets relatively inexpensively and cut them with a handsaw or, better yet, a bandsaw. Better still, if you have access to one, a laser cutter allows you to create perfect robots to exact, computer-given dimensions and designs, as you can see in Figure 3-4. Check your local area to see if there is a makerspace/hackerspace that has a communal laser cutter you could use.

FIGURE 3-4
The base of a custom laser-cut robot chassis that I designed and made in a makerspace in Cambridge, UK

You could also work with plastic using a 3D printer. 3D printers work by heating up and extruding layers of plastic to build an object layer by layer. You can design your chassis in computer-assisted design (CAD) software on a computer, or you can even download and print other people's designs from the internet. 3D printers are becoming much more commonplace, so you may be able to find one at a local library, makerspace, or hackerspace, or you may even know someone who owns one.

If you don't want to make your own chassis, you can buy a variety of basic kits online. Most of these are either one or two layers of acrylic that can be fastened together with screws and pillars (see the example in Figure 3-5). You can find a premade chassis by searching "robot chassis" on the internet. eBay usually has some of the best deals.

Motors

Without motors, your robot can't move. And a robot that can't move isn't much of a robot! Let's look at a few motor basics so you know what kind of motors to use when.

What Is a Motor?

An electric motor converts electrical energy into mechanical energy. Motors come in many different sizes and designs, and a wide range of prices. For this book, you'll want to get two (one for each wheel) of the cheapest type: *DC motors*. These are the most common kind of motor and can be found in everything from trains to desktop fans.

DC stands for *direct current*, which means the electric current must flow through the motor in one direction for it to work. Alternatives to DC motors—such as AC (alternating current) motors and stepper motors—tend to be more expensive, harder to use, and more complicated electronics-wise, so we'll stick with DC for now.

The super-simple motors we'll use have two terminals. A terminal is just a point where electricity can enter and leave a component. When you apply a voltage over these terminals, the motor shaft will spin. If you change the direction of the voltage (sometimes referred to as *reversing* the voltage), the motor will spin the other way.

You can spend ever-increasing amounts of money to get ever better motors. It's a good idea to start off *really cheap* to get the basics right before investing in more expensive equipment.

Motor Options

For your first robot build, get a pair of geared *brushed* (not brushless) DC motors. In both brushed and brushless motors, electricity creates electromagnetic forces that are responsible for spinning the motor shaft; however, brushless motors are more complicated and require additional expensive circuitry to function. Brushless motors are usually used for more serious tasks, like remote-controlled model airplanes and drones.

You can pick up a pair of brushed DC motors from any of the usual online retailers for less than $10 (see "Where Should You Buy Parts?" on page xix for a list of retailers). Before buying, ensure you take a look at the motor's specification. You'll need to consider a few of the following factors.

I advise getting motors rated between 5 V and 9 V. If your motors' voltage requirement is too high, you'll struggle to power them and will likely need more batteries. If the voltage of your motors is too low, the voltage provided by your power source (see the next section) may cause overheating and damage!

The amount of current a motor draws is also very important. The more current needed, the quicker your batteries will deplete and the harder the motor will be to control. On the other hand, less current equates to less powerful motors that may struggle to haul your robot around. Try to ensure that your motors are rated at no more than 500 mA each. Too much current may also overload the motor controller we'll be using later.

A cheap motor will usually have a high number of *revolutions per minute (RPM)*, spinning around 1,000 to 3,000 times in a single minute, which is far too fast for a small robot. At these RPMs, each motor will create very little *torque*. Torque is the driving force, and the smaller it is, the more your robot will struggle to move along a surface. To fix this, your motors need a gearbox to bring the RPM down and increase the torque. The ratio of the original RPM to the new geared, lower RPM is called its *reduction rate*, and 48:1 is a decent reduction rate for this project. When searching for your supplies, make sure you're looking for motors that have a small gearbox preinstalled by using search terms like "geared hobby motors."

Motors are not very useful without wheels. Most motors will have a *shaft* you can fit a tire onto that will revolve when the motor runs and consequently turn the tires. If you buy motors without tires, you'll have to source ones that will fit the shafts of the motor you have purchased, which can be hard. I recommend buying motors that come with tires. Grippy rubber wheels will provide you with good control. Alternatively, you could always try making your own wheels, especially if you have access to precision equipment like a 3D printer.

The motors that I use, shown in Figure 3-6, fit all of these criteria. These are common hobby motors that have a gearbox and come with their own set of tires. They can also run off voltages anywhere between 3 V and 9 V and draw around 100 mA of current each. You can pick up a pair of these for as little as $5. I sourced mine from eBay by searching for "robot motor with tire," though they're widely available from many other places.

FIGURE 3-6

My robot's motors and tires

Most DC motors will come without wires attached to the two terminals. If this is the case with your motors, you'll have to *solder* wires onto them. Soldering is the process of electrically joining two components by melting a filler metal called *solder* between them. This can seem very daunting, but don't worry—it's really easy and a super important skill! Take a look at "How to Solder" on page 204 for more information.

If you really want to avoid soldering, you may be able to find motors online that come with wires already soldered and attached to the terminals. Just keep in mind that these may be hard to source and will probably command a higher price!

Batteries

To make sure our robots can move around by themselves, we'll power them with batteries so we don't have trailing power cables. A *battery* is an electrochemical device that stores energy.

For this book, you'll power your robots with AA batteries. Not only are these common and cheap, they are also safe and easy to use. A single AA battery will usually provide between 1.2 V and 1.5 V of power, and they can be chained together to provide larger voltages. The number of AA batteries you need depends on the voltage of your motors. Six AA batteries chained together in a battery holder will provide an output voltage between 7 V and 9 V, which should be the right amount for your robot.

AA batteries come in two variations: *primary* (nonrechargeable) and rechargeable. Primary batteries are cheaper but can be used only once and then need to be responsibly thrown away. This means that, over time, the cost of using primary batteries will creep up. I recommend investing in some high-quality rechargeable batteries and a decent charger. The initial expense can be higher ($20–$30), but you'll save money in the long run and be more environmentally friendly in the process! I use Panasonic's Eneloop rechargeable AA batteries, shown in Figure 3-7. Bear in mind that rechargeable batteries often have a slightly lower voltage than primary batteries (for AA, usually 1.2 V rather than 1.5 V).

To store and connect these batteries, you'll need a *battery holder*. This will both hold your batteries in place and connect the terminals of your batteries together so that you have just one positive and one negative wire to connect to get power out of all of them. These are widely available online. I picked up the six AA battery holder with an on-off switch shown in Figure 3-7 from eBay for $1.

FIGURE 3-7

My battery holder
with six AA Eneloop
rechargeable batteries

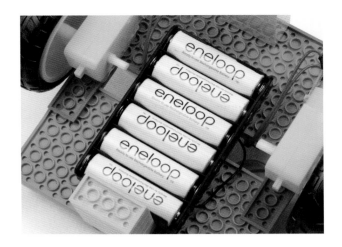

Voltage Regulator

While your motors will happily run off the 7 V to 9 V your batteries provide, your Raspberry Pi will *certainly* not. Your Raspberry Pi operates on strictly 5 V (with a tolerance of 0.1 volts on either side). If you provide less voltage, your Pi won't boot. If you provide any more, you'll break your Pi! Providing too much voltage will allow too large of a current to flow through your Pi's internal components and the Pi will blow up (which, in reality, means a small plume of blue smoke will come from the processor, which will be irreversibly damaged).

To avoid blowing your Pi up, you can use a simple voltage-regulating component called a *step-down buck converter* to turn the 7 V to 9 V into the 5 V your Pi needs. A step-down buck converter reduces an input voltage to a desired output voltage. We'll use a converter based around the LM2596 chip, which has been neatly arranged into an easy-to-use module board, as shown in Figure 3-8. This module can take a voltage input between 4 V and 40 V and reduce it to anything between 1 V and 35 V. I'll show you how to set yours to 5 V for your Raspberry Pi later in the chapter.

You can pick up an LM2596 module, or another similar buck converter, online for a few dollars. If you decide to go for a different model, ensure that it is able to output *at least 2 A of current continuously*. This information should be available in the listing of the product or in its *datasheet*, a reference that details the technical characteristics of a component. Also note that you should purchase a buck converter that uses screw terminals for the inputs and outputs—this makes things easier and saves you from more potential soldering.

FIGURE 3-8
An LM2596 buck converter module

Motor Controller

Your DC motors will draw up to 500 mA each. For comparison, the Pi's GPIO pins can provide only 20 to 50 mA in total. This means that your motors will need to be powered directly from your separate battery pack. This isn't a problem, but it does mean the Raspberry Pi isn't directly connected to the motors, so you'll need a *motor controller* to interface between the motor, its power source, and your Raspberry Pi. A motor controller will allow you to use your Pi to turn your motors on or off, and to control their speed.

There are many different motor controllers available, in a wide variety of packages. You can get driver chips, module boards, or even *HATs* (official Raspberry Pi *Hardware-Attached-on-Top*). Each option has advantages and disadvantages.

In this book, I'll use an *integrated chip (IC)* called the L293D. A breadboard-friendly IC, like my L293D in Figure 3-9, is a small black box that contains a collection of miniaturized electronic components, like resistors, transistors, and capacitors. You can insert the legs of the IC into the breadboard and wire them up to provide extra circuit functionality. The L293D gives you complete control over up to two independent motors—perfect for your first robot!

FIGURE 3-9
An L293D motor controller chip

For more specific details on the vast capabilities of the L293D motor driving chip, just search online for its datasheet. You can pick up an L293D for less than $4 online.

Recommended Tools

Over the course of the robot-building process and the rest of the book, you'll need some basic tools. This is a list of all the tools you'll need, as well as a few optional tools that aren't strictly necessary but might help you:

- Variety of screwdrivers
- Hot glue gun
- Multimeter
- Soldering iron

ASSEMBLING YOUR ROBOT

Once you have the components for your robot, you can start assembling and wiring it up! If you bought the same components as the ones I listed earlier, you can follow my instructions exactly. If you bought or made slightly different components, you may have to get a bit creative, but the following instructions should be an excellent guide no matter what parts you have.

Making the Chassis

As mentioned, I'm making my robot chassis out of LEGO pieces, so that's what these instructions will use.

You can make a chassis out of LEGO pieces in an infinite number of ways with an infinite number of brick combinations. My very simple chassis uses the following parts, shown in Figure 3-10:

- Two 8×16 plates
- Four 2×8 plates

If you want to purchase these exact LEGO pieces and follow this process precisely, you can use the "Pick A Brick" service at *https://shop.lego.com/en-US/Pick-a-Brick/*. Here you can order individual pieces for your own custom creations by searching for the pieces using their element IDs. You'll need:

- Two 8×16 plates with ID 4610353
- Four 2×8 plates with ID 303428

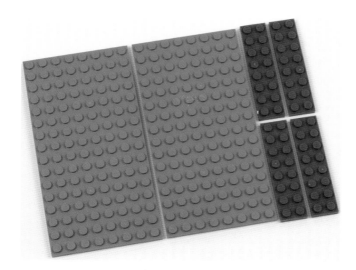

FIGURE 3-10
The LEGO pieces I'm using
to build my chassis

Plug these ID numbers into the Element ID search box and you should find them.

It is worth noting that, according to the LEGO website, you may have to wait up to 10 business days for your parts to arrive! If you choose to make your chassis from LEGO pieces, you'll need some other bricks later on in the building process, so I suggest ordering all the bricks you need at once.

The 8×16 plates are perfectly sized to mount a Raspberry Pi and a breadboard each. I've made mine with a gap between the two LEGO plates that makes it easy to feed the wires through neatly.

To follow my design, place your 8×16 plates two LEGO studs apart, and attach two 2×8 plates on the top to fasten the larger plates together. Fasten the other two 2×8 pieces on the underside to make it extra sturdy, as shown in Figure 3-11.

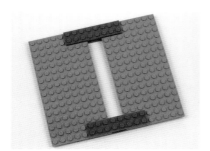

FIGURE 3-11
The completed LEGO chassis: on the right is a close-up of the joints made with the 2×8 plates

This sandwiches the plates together to create a sturdy and stable platform.

Attaching the Motors

With your chassis constructed, you can attach and wire up the various components, starting with the motors. Before you attach the motors to the body of your robot, ensure that they have wires soldered to their terminals! Again, see "How to Solder" on page 204 to learn how to do this.

You'll mount your motors to the bottom of your chassis, with the wheels as central to the body as possible to give your robot a small turning circle—meaning it can make tighter turns. I'll give you a few options for securing the motors. Whichever option you choose, make sure that your motors are aligned and as parallel as possible. Nonparallel motors will cause conflict in the direction your robot travels and may prevent your robot from moving in a straight line.

Permanent Option: Glue

Gluing with either a hot-glue gun or superglue will securely and permanently bond your motors to your chassis. This is my preferred way of attaching the motors to the chassis, but think twice before you glue, since it's permanent! Make sure you're completely satisfied with the motor positions before you commit.

A neat layer of glue as shown in Figure 3-12 ensures that your motor joints are secure and do not wiggle.

FIGURE 3-12

Hot glue used to attach the motors

Less Permanent Option: Velcro or Screws

Traditional hook-and-loop fastener material (more commonly known as Velcro) is a great way of fixing things together in a nonpermanent way, with no mess and no tools required. My favorite brand is

3M Dual Lock, which is an incredibly strong hook-and-loop pad, though slightly more expensive than other solutions at around $15 per meter. Hook-and-loop fasteners usually have an adhesive back so you can just cut off as much as you need, peel the covering away, and stick it onto your chassis. If you find this isn't quite strong enough, or has too much flex for mounting your motors, I recommend adding a little bit of glue around the edges of the Velcro/ Dual Lock.

Alternatively, another less permanent option is using screws. If you've made a wood, laser-cut plastic, or 3D-printed base, you could screw your motors to the chassis. Some motors even come with brackets and screws for this exact purpose. This is a secure way of doing things, and you can always unscrew your motors to use in a different project further down the line.

Even Less Permanent Option: Sticky Tape

Double-sided sticky tape is useful for securing parts together easily and without fuss, similar to Velcro, although this is the weakest option.

Stabilizing the Robot

After you've attached the motors to the bottom of your chassis, feed the two wires of each motor through the gap up onto the top of your platform, as shown in Figure 3-13.

FIGURE 3-13

The chassis with motors connected, and wires threaded through the gap

With your two motors attached in the center of the chassis of your robot, you may notice that the platform is unstable and resting on its heaviest side. Don't worry: you can easily fix this!

Stabilizers will prevent your robot from rocking and allow for smooth movement. You need only one stabilizer on either side, and this can be made out of anything. Because I'm using a LEGO chassis, I've constructed mine out of LEGO bricks! I put two five-deep 2×4 brick posts on the front and the back underside of my robot; these will stabilize the weight but won't touch the ground, so they shouldn't obstruct movement when running on a smooth surface. These guides have eliminated the rocking entirely; see Figure 3-14.

FIGURE 3-14

The chassis with LEGO stabilizers on the front and back underside

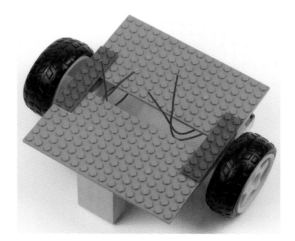

If you're also using LEGO parts, you may need more or fewer bricks—just make sure your stabilizers span from the bottom of the chassis to just above the floor. Be careful not to make the stabilizers too long, as they may hinder your robot's movement or suspend its motors above the ground completely!

If you don't want to make your stabilizers out of LEGO bricks, you can use any other material (pieces of plastic, wood, etc.) that fits between the chassis and the floor, and that you can firmly attach to your chassis.

A common option is to use one or two caster wheels or balls; these are nonmotorized wheels or balls used solely to balance the robot. You should be able to find these online in the dimensions you need—just measure from the chassis to the floor when the chassis base is level—and you'll often find they have screw mounts, which makes them a particularly good option if your chassis is made out of wood or plastic.

Many of the robot platforms you can buy online come as complete kits with the motors, wheels, and caster wheel. This can be a great way to save time, especially as all the holes have been drilled for you.

Attaching the Batteries

Now that you have your motors attached to your robot, you can fix your battery holder and batteries in place. Figuring out where to mount the batteries involves two vital considerations:

Space Where is there enough space for your battery holder? Will the wires from the battery holder reach the top of your chassis? Will your battery holder obstruct the movement of the robot in any way?

Access At some point you'll need to remove the batteries and either replace or recharge them, so you need to make sure you have easy access to your holder. Don't make the mistake of permanently sticking your battery holder on, only to not be able to get at the batteries inside!

I have mounted my battery pack on the underside of my LEGO chassis and between the motors, as shown in Figure 3-15. This is a good option because it leaves the top free to mount the components you need to access more regularly, like your Pi, while still giving you easy access to the batteries to replace them. I've mounted the holder using some trusty 3M Dual Lock. Try to center the battery holder to keep the weight balanced.

WARNING

Ensure that the positive and negative wires of your battery holder don't accidentally touch while the batteries are in place. If they do, they'll create a short circuit, which will cause rapid heating and potentially even damage your batteries.

FIGURE 3-15

The underside of the robot with a battery holder attached

Once you've fixed your battery holder into place, insert some new or fully charged batteries into it and thread the wires from the holder up to the top of the chassis.

Mounting the Raspberry Pi, Breadboard, and Buck Converter

After you've set up your motors and batteries, flip your robot the right way up to mount the key electronics: your Raspberry Pi, breadboard, and buck converter.

The 8×16 LEGO plates are perfectly sized to fit a Pi and a breadboard each. As you can see in Figure 3-16, I have mounted my Raspberry Pi on one plate and the breadboard on the other. In the bottom-right corner, you can also see the LM2596 buck converter. This is my recommendation for positioning your parts.

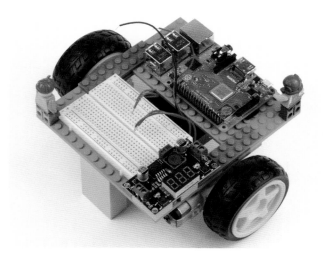

Since you'll likely want to use your Pi again, I'd recommend a nonpermanent adhesive option. I used sticky tack to delicately and nonpermanently attach my Pi. Most breadboards come equipped with an adhesive back, so I simply peeled off the cover and used the adhesive to securely fasten my breadboard to the LEGO plate. If your breadboard has no adhesive underside, then I recommend using sticky tack for that, too.

Depending on what your chassis is made out of, you may choose a different way of fastening your electronics. The latest Raspberry Pi and its previous models come with screw holes, so you could use a set of small screws to fasten your Pi to a wood or plastic chassis.

I attached the LM2596 buck converter module next to my breadboard using more sticky tack, suspended on two 2×2 LEGO bricks. When you do this, ensure that the screw terminals of your voltage regulator are easily accessible and not blocked by your breadboard or anything else, because you'll need to connect wires to these later.

Wiring Up Power to the Raspberry Pi

Before you can wire up your motors, you need to set up the power going to the brain of your robot: the Raspberry Pi.

As mentioned, it's crucial that your Pi gets 5 V (±0.1 V), and not the 7 V or 9 V (or other voltage) that your batteries provide. This means you'll need to adjust your buck converter to output precisely 5 V.

The best way to do this is with the help of a tool called a *multimeter*, shown in Figure 3-17. You should be able to pick up a multimeter for as little as $10 online.

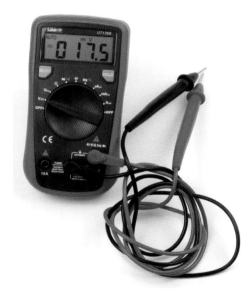

FIGURE 3-17

A multimeter, set to read voltage

A multimeter is an electronics measuring instrument. A normal multimeter can measure voltage, current, and resistance. We're interested in measuring the voltage.

Some voltage converters, like the LM2596 module I am using, do have a small LED screen that displays a voltage readout. While in theory this eliminates the need for a multimeter in this case, it is still good practice to use a multimeter to double-check the output voltage. A multimeter will come in handy in the future too, so it's a wise investment.

We're going to connect the power through the GPIO pins on your Pi. Follow these instructions carefully, taking your time, and your Pi will have power without blowing up! But don't worry too much: this way of powering a Pi has been tried and tested many times and is totally safe when done right!

Setting Up the Converter

Start by setting up your buck converter using the following instructions—you won't connect it to the Raspberry Pi just yet. Flip to the circuit diagram in Figure 3-21 to see a final diagram of the buck converter setup. Check out Figure 3-19 for a close-up image of the buck converter too.

1. Connect both wires from your battery pack into *separate* columns and rows of your breadboard. Most breadboards will have two rows running along each edge that are lined with red and blue or have a + or – sign at the end; these rows are known as *power rails*. To help keep track of where your negative and positive power lines are connected, connect the black (negative) wire from your battery pack into one side of your breadboard's blue – power rail, and the red (positive) wire to that side's red + power rail. Check out Figure 2-4 on page 34 for a diagram of these power rails.

BATTERY PACK	BREADBOARD
Red, positive wire	+ power rail
Black, negative wire	– power rail

Now we'll connect the LM2596 module. Most LM2596 buck converter modules come equipped with *Philips-head screw terminals*. To connect a wire to these terminals, simply unscrew the terminal you want, place a wire in the hole at the front, and then tighten the screw to make a firm connection.

2. Take one red and one black M-M jumper wire and insert them into the positive and negative ground rails, respectively. Ensure that the ends of the wires do not touch during this. Now unscrew the screw in the terminal labeled VIN (for voltage in) of your LM2596 module, insert the red positive wire, and then tighten the screw terminal. Then insert the black negative wire into the ground terminal labeled GND (for Ground) on the same side of your LM2596 module, as shown in Figure 3-18. Make sure that you get the positive and ground wires the right way around!

FIGURE 3-18

The 7 V–9 V and ground
lines running into the
LM2596 buck converter

LM2596	CONNECTION
VIN	Red wire, + power rail
GND	Black wire, – power rail

3. Next, turn on your multimeter and set it to read voltage—on
    most multimeters you do this by turning the dial in the middle
    to *V*. Then insert the multimeter's red positive lead into the VOUT
    (voltage out) terminal of your buck converter, and insert the mul-
    timeter's black negative lead into the GND terminal on the same
    side (see Figure 3-19). You should see a reading of the output
    voltage of your buck converter appear on the multimeter. Don't
    worry about the value of the reading just yet!

FIGURE 3-19

A close-up of the buck
converter

4. With the multimeter connected so you can see how the voltage changes, use a screwdriver to twist the *adjustment screw* on the top of your buck converter (check back to Figure 3-19). Insert your screwdriver, and as you twist either left or right your voltage reading should change. Work out the correct direction you need to twist the adjustment screw, and then keep twisting the screw until your multimeter reads 5 V (±0.1 V), as shown in Figure 3-20. This requires a little trial and error!

FIGURE 3-20

The multimeter reading 5 V from the output of the buck converter

5. Once you are satisfied that your multimeter is reading 5 V (±0.1 V), unplug it from the output of your buck converter and turn it off.

Wiring Up the Converter

Now that you have adjusted the output voltage of your buck converter, you can wire it up to your Raspberry Pi. Make sure your batteries are off, or even take the batteries out of your battery holder, before wiring up the converter.

NOTE

Slightly overpowering your Raspberry Pi with a voltage of no more than 5.1 V can be beneficial. It can prevent your Pi from cutting out due to the small voltage drop from the batteries that can occur when the motors are running, for example.

6. Grab two M-F jumper wires—the convention is to use one red for positive and one black for negative (Ground). You'll connect the male ends of the jumper wires to each terminal of your converter's *output*. As usual, connect red to the positive output (VOUT) and black to the negative output (GND) of your LM2596 module. Then, connect the red wire to physical pin 2 on your Raspberry Pi (+5 V). Finally, connect the ground wire to physical pin 6 of your Pi (Ground). See Figure 3-21 for a breadboard diagram and image showing how your Pi should look. Also refer to "Raspberry Pi GPIO Diagram" on page 200.

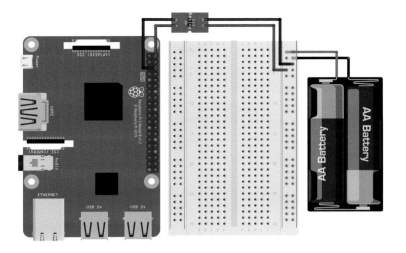

FIGURE 3-21

A breadboard diagram of the buck converter circuit (top); the wired-up LM2596 converter (bottom)

WARNING

Before you proceed, double-check that you are happy with your wiring and the voltage your buck converter is providing. If you are ever in doubt, unplug your Pi and check the voltage again using your multimeter before you turn the power on. It never hurts to check multiple times!

Good work—you've finished wiring your buck converter up to the Pi's GPIO pins! Now flick the power switch of your battery holder on (or reinsert the batteries), and you should see your Raspberry Pi spring to life just as if you had connected power to the micro USB port.

Wiring Up the Motors

The final stage of your base robot build is to wire up your motors to your motor controller, and then wire your motor controller to your Pi. Bear in mind that this process may be different if you are using a different motor controller than I am.

FIGURE 3-22

The L293D inserted into my breadboard

Connecting the L293D Motor Controller

The following instructions are for the L293D motor controller chip that I'm using:

1. Insert your L293D firmly into your breadboard, ensuring that all of the legs are in separate rows and are not connected to other components. To do this, position the chip so it straddles the gap running down the middle of your breadboard, as shown in Figure 3-22. The legs of ICs like the L293D often are not quite straight, so you may struggle to fit it into your board. If this is the case, take your L293D and press each side lightly against a flat surface to bend the legs into right angles.

2. The L293D has 16 legs, none of which are labeled, so you'll need to refer to the pin diagram in Figure 3-23 to connect the motor controller. You can also usually find this information in the chip's datasheet. ICs are numbered counterclockwise starting from the top-left pin. Pin 1 on your L293D is to the left of the notch at the top of the chip. Sometimes there's a dot on the chip next to pin 1 too.

3. The L293D needs its own source of power to function, which we can provide through the Raspberry Pi's 5 V pin. We're already using the first 5 V pin, so use an M-F jumper wire to connect the second 5 V pin (physical pin 4) of your Pi to the red power rail on the edge of your breadboard that you're *not already using*.

Use another M-F jumper wire to connect a ground pin of your Pi (physical pin 9) to the blue power rail on the same side; see Figure 3-24. *Make sure you're not connecting anything to the power rail that you previously wired up!*

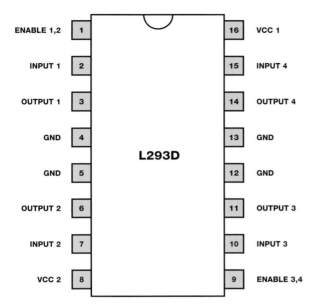

FIGURE 3-23

A pinout diagram of the L293D

4. Now we'll connect the grounds of the two power rails up. Connect the two ground rails on your breadboard by running a black M-M wire from one blue power rail to the other. This connects your Pi's ground rail to your battery's ground rail, as shown in Figure 3-24. We refer to connected grounds as the *common ground*.

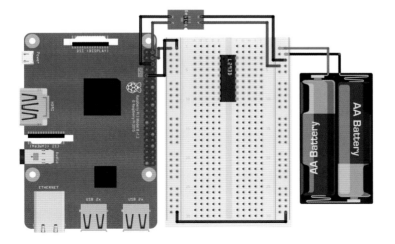

FIGURE 3-24

The L293D IC inserted into the breadboard with the Pi's power rails connected

5. Next use a wire to connect pin 16 (VCC 1) of your L293D to the 5 V positive power rail (the one connected to your Pi's 5 V pin). This is the chip's power supply. Once you have done this, connect the ground pins (GND) found on pins 4, 5, 12, and 13 of your L293D to the common ground. See the breadboard diagram in Figure 3-25 for guidance.

FIGURE 3-25

5 V power connected to the L293D and all four grounds wired up

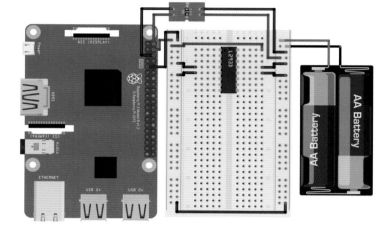

6. Now it's time to connect the power for the motors, which will come directly out of your battery pack. Wire the positive +7 V to 9 V power rail connected to your battery pack to pin 8 (VCC 2) on your L293D. Instead of being the chip's power supply, this time this is the motor's power supply. Figure 3-26 shows this connection.

7. The L293D has two *Enable* pins. If these aren't *on* (in other words, if they are not receiving a high voltage), then any motor wired up to your L293D won't respond to commands. To turn the Enable pins permanently on, connect pin 1 (Enable 1,2) and pin 9 (Enable 3,4) of your L293D to the positive power rail connected to 5 V on the Pi. I have used white wires in Figure 3-26 to show this step.

NOTE

The length of your jumper wires can sometimes make them ungainly and cause your wiring to be confusing. If you got a variety pack of breadboard wires, you can use the shorter ones here to make the board neater. Alternatively, buy some single-core wire and cut and strip your own pieces to size.

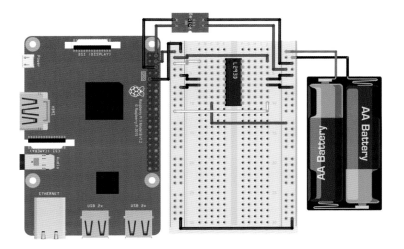

FIGURE 3-26
Connecting the L293D's
Enable pins to a high
voltage by connecting them
to the Pi's 5 V power rail

Connecting the Motors

You can now wire up your motors to the L293D.

8. Take one motor and connect one of its wires to pin 3 (Output 1) of
    your L293D chip, and the motor's other wire to pin 6 (Output 2).
    Then take your second motor and connect the first of its two wires
    to pin 11 (Output 3) and the second to pin 14 (Output 4), as shown
    in Figure 3-27.

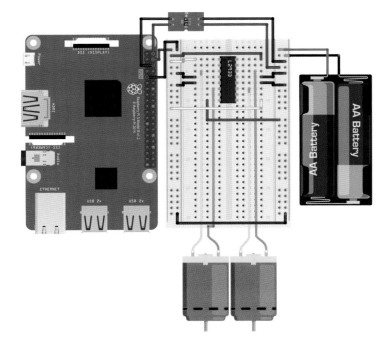

FIGURE 3-27
Both motors wired up to
the outputs of the L293D

Finishing Up the Wiring

The last stage of the wiring process is to connect your Pi's GPIO pins to the inputs for the L293D. Each motor requires two GPIO pins to function (I'll cover how this works in the next chapter). This means that, in total, you need four GPIO pins to drive two motors.

9. Use a M-F jumper wire to connect your Pi's physical pin 11 (BCM 17) to pin 2 (Input 1) on your L293D. Then use another jumper wire to connect your Pi's physical pin 12 (BCM 18) to pin 7 (Input 2) on your L293D. You have now successfully wired up one motor. At this stage, your breadboard should look like Figure 3-28, with the most recent additions being the pink wires.

FIGURE 3-28

First motor's input control pins wired from the Pi to the L293D

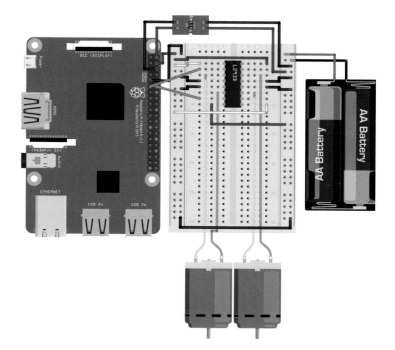

10. Finally, you must wire up the GPIO pins for the second motor. Do this by connecting your Pi's physical pin 13 (BCM 27) to pin 10 (Input 3) of your L293D with another jumper wire. Then connect your Pi's physical pin 15 (BCM 22) to pin 15 (Input 4) of your L293D.

Congratulations! You have finished wiring up your motors and motor controller. Your breadboard should look like the diagram in Figure 3-29, which shows the last wires in purple. In addition, it shows a photo of my robot all wired up.

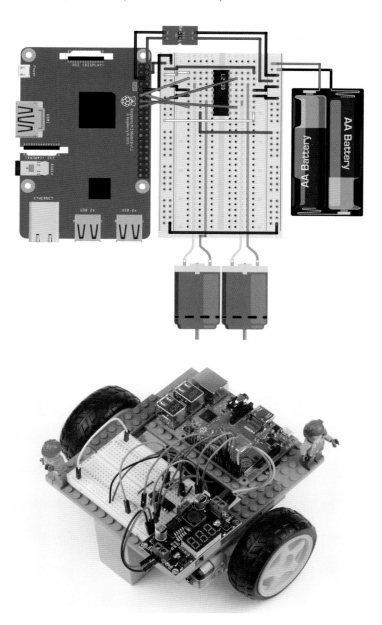

FIGURE 3-29

The completed circuit, with both motor's inputs wired from Pi to L293D (top); the completed robot! (bottom)

You have the physical parts of your first robot finished. The next step is to program it to do cool stuff. As it is, you can program your robot to move around like a remote-controlled car. In the next chapter, we'll add some code so you can do exactly that!

SUMMARY

You have done *a lot* in this chapter! We have discussed why two-wheeled robots are great, their various elements, and the options you have for building them. I've shown you how to wire up your Pi, connect power, and attach your motors to build your very own robot from scratch.

With all of this fantastic groundwork in place, in the following chapter, I'll guide you through the next exciting step: programming your robot and making it move!

As with any wiring, a little perseverance pays off! Before you proceed to the next chapter, quickly read through the instructions again to double-check that you have no short circuits and everything is connected as it is supposed to be. That way, you should have no problems getting your robot up and running in the next chapter.

4
MAKING YOUR ROBOT MOVE

AT THIS STAGE, YOU HAVE A SWEET-LOOKING RASPBERRY PI ROBOT THAT DOESN'T DO ANYTHING... YET! TO UNLOCK THE CAPABILITIES OF ALL THE HARDWARE YOU JUST WIRED UP, YOU'LL HAVE TO SINK YOUR TEETH INTO SOME MORE PROGRAMMING.

In this chapter, I'll show you how to use the Python programming language to make your robot move. We'll cover basic movement, making your robot remote-controlled, and varying its motor speed.

THE PARTS LIST

Most of this chapter will be about coding the robot, but to enable remote control you'll need a couple of parts later:

- Nintendo Wii remote
- Bluetooth dongle if you're using a Pi older than a Model 3 or Zero W

UNDERSTANDING THE H-BRIDGE

Most single-motor controllers are based around an electronics concept called an *H-bridge*. The L293D motor driver chip we're using contains two H-bridges, permitting you to control the two motors of your robot through a single chip.

An H-bridge is an electronic circuit that allows a voltage to be applied across a load, usually a motor, in either direction. For the purposes of robotics, this means that an H-bridge circuit can drive a motor both *forward* and *backward*.

A single H-bridge is made of four electronic switches, built from transistors, arranged like S1, S2, S3, and S4 in Figure 4-1. By manipulating these electronic switches, an H-bridge controls the forward and backward voltage flow of a single motor.

FIGURE 4-1

A single H-bridge circuit

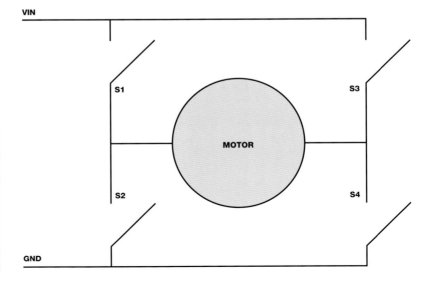

When all the switches are open, no voltage is applied to the motor and it doesn't move. When only S1 and S4 are closed, there is a flow of current in one direction through the motor, making it spin. When only S3 and S2 are closed, a current flows in the opposite direction, making the motor spin the other way.

The design of the L293D means that we can't close S1 and S2 at the same time. This is fortunate, as doing so would short-circuit the power, causing damage! The same is true of S3 and S4.

The L293D abstracts this one step further and requires only two inputs for one motor (four inputs for a pair of motors, like you wired up in Chapter 3). The behavior of the motor depends on which inputs are high and which are low (1 or 0, respectively). Table 4-1 summarizes the different input options for the control of one motor.

INPUT 1	INPUT 2	MOTOR BEHAVIOR
0	0	Motor off
0	1	Motor rotates in one direction
1	0	Motor rotates in other direction
1	1	Motor off

TABLE 4-1
Motor Behavior Based on Inputs

We'll use the GPIO Zero Python library to interface with the Pi's GPIO pins and motor controller. There are several functions in the library for controlling basic movement, so you won't have to worry about turning specific GPIO pins on and off yourself.

FIRST MOVEMENT

Now for the most exciting step of your robotics journey yet: moving your robot! You'll eventually make your robot entirely remote-controlled and even able to follow your instructions, but before that let's master some basic motor functionality. You'll start by programming your robot to move along a predefined route.

Programming Your Robot with a Predefined Route

Boot up your Raspberry Pi on your robot and log in over SSH. While your robot is stationary and being programmed, it is best to disconnect your batteries and power your Pi from a micro USB cable connected to a wall outlet. This will save your batteries for when they are really needed.

From the terminal, navigate from your home directory into the folder you are using to store your code. For me, I'll navigate into my *robot* projects folder like so:

```
pi@raspberrypi:~ $ cd robot
```

Next, create a new Python program and edit it in the Nano text editor with the following command; I have called my program *first_move.py*:

```
pi@raspberrypi:~/robot $ nano first_move.py
```

Now you need to come up with a predefined route to program! With the DC motors we're using, you *can't* rotate them a specific distance or number of steps, but you *can* power them on and off for a certain amount of time. This means that any path will be a rough approximation of where you want your robot to go rather than a precise plan.

To start, let's keep things simple and make your robot drive around in a square, with a route like the one shown in Figure 4-2.

FIGURE 4-2

The robot's planned route

In your *first_move.py* file, enter the code in Listing 4-1 to program a square route.

LISTING 4-1

Programming your robot to move in a square

```
import gpiozero
import time
```

❶ `robot = gpiozero.Robot(left=(17,18), right=(27,22))`

❷ `for i in range(4):`
   ❸ `robot.forward()`
   ❹ `time.sleep(0.5)`
   ❺ `robot.right()`
   ❻ `time.sleep(0.25)`

The program starts by importing familiar Python libraries: `gpiozero` and `time`. Then you create a variable called `robot` ❶, to which you assign a `Robot` object from the GPIO Zero library.

Objects in Python are a way of holding variables (pieces of information) and functions (predefined sets of instructions that perform tasks) in a single entity. This means that when we assign an object to a variable, that variable then has a range of predefined things that it knows and can do. An object gets these capabilities from its *class*. Each class has its own functions (called *methods*) and variables (called *attributes*). These are advanced features of Python and you don't have to worry about them too much at this stage. Just know that we're using some predefined classes from Python libraries, like GPIO Zero, to make it easier for us.

The GPIO Zero library has an inbuilt `Robot` class that features a variety of functions for moving a two-wheeled robot in different directions. Notice the two sets of values in the parentheses assigned to `left` and `right` ❶. These represent the input pins of the L293D you have wired up. If you followed my exact wiring from Chapter 3, then the four GPIO pins should be: 17, 18 and 27, 22.

This program also uses a new type of loop called a `for` loop ❷. In Chapter 2, while making LEDs flash on and off and getting inputs from buttons, you used a `while` loop. A `while` loop keeps repeating its contents indefinitely while a certain condition is met, but a `for` loop repeats a block of code a *fixed* number of times. The syntax of this loop, `for i in range(4):`, means "do the following four times."

The `for` loop commands your robot to start going forward ❸ and then wait for half a second ❹ to allow some time for the robot to move. The result is that both motors move in a single direction (forward) for half a second.

You then instruct your robot to turn right ❺ and wait for a quarter of a second as this happens ❻. By telling the robot to turn right, you replace the forward command issued half a second ago with a new command for the motors.

Once this has been executed once, the "go forward, then turn right" process starts again and continues for a total of *four* times. You are trying to make your robot go in a square, and squares have four sides, hence the specific repetition.

Once you've finished writing your program, exit Nano by pressing CTRL-X and save your work like usual. Next, we'll run the program to make the robot move!

The GPIO Zero Robot class has commands for all directions and basic functionality, summarized in Table 4-2.

TABLE 4-2

The Robot Class
Commands

COMMAND	FUNCTIONALITY
robot.forward()	Run both motors forward.
robot.backward()	Run both motors backward.
robot.left()	Run the right motor forward and the left motor backward.
robot.right()	Run the left motor forward and the right motor backward.
robot.reverse()	Reverse the robot's current motor directions. For example: if going forward, go backward. If going left, go right. This is *not* the same as going backward!
robot.stop()	Stop both motors.

Running Your Program: Make Your Robot Move

Before you execute your program, ensure your robot is disconnected from the wall power outlet and the batteries are connected and turned on. You should also place your robot on a relatively large, flat surface clear of obstacles and hazards. Rough surfaces, like carpets, may cause your robot to become stuck or struggle to move. Try to avoid this, as struggling motors draw more current, and when their movement is completely blocked (or *stalled*) you might even damage your electronics! The flatter the surface, the better your robot will run.

It is also a good idea to be in a position to "catch" your robot in case either it or something/someone is in peril. It may try to go down the stairs, for example, or the cat may be in the way.

To run your program, wirelessly access the terminal of your Pi using SSH and enter:

```
pi@raspberrypi:~/robot $ python3 first_move.py
```

Your robot should burst into life and start to move. If all has gone well, it will move on a square-based path and then come to a stop, and your program will end by itself. If you need to stop your robot at any point, press CTRL-C on your keyboard to kill the motors immediately.

TROUBLESHOOTING GUIDE: ROBOT NOT WORKING PROPERLY?

If your robot isn't functioning as it should be, don't worry. Usually malfunctions fall into some common categories and should be easy to fix! The following quick guide will help you resolve most issues you might have.

Robot Moving Erratically

The most common problem after you execute the *first_move.py* program is that your robot moves, but not in the right pattern. Instead of going forward, it goes backward; or instead of turning right, it turns left. You may even find that it just spins on the spot!

This behavior can be easily fixed. As we discussed, DC motors have two terminals with no particular polarity. This means that if you change the direction of current flowing through the motor, the motor spins the other way. Consequently, if one or both of your motors is going in the opposite direction of your commands, you can swap the wires connected to the output pins of your motor controller to reverse this. For example, swap the wires connected to Output 1 with Output 2 of your L293D. Refer to Chapter 3 for guidance and relevant diagrams.

(continued)

Motors Not Moving

If your program successfully executes, but your robot's wheels don't move or only one motor starts to move, then you could have an issue related to your wiring. Go back to the previous chapter and check that you've connected everything as per the instructions. Ensure the connections to the motors are solid and that none of the wires have become loose. If you're convinced that you've wired everything correctly, check whether your batteries are charged and that they can provide enough power for your specific motors.

If your Raspberry Pi crashes when the motors start to turn, you most likely have a power issue. Check how you have set up your buck converter. If you are using a different converter than mine, you may run into problems. Go back a chapter for guidance and recommendations.

Robot Moving Very Slowly

A slow robot is usually a sign that not enough power is being provided to the motors. Check the voltage requirements of your motors and make sure you're supplying them with what they need. Often motors will accept a range of voltages—for example, from 3 V to 9 V. If your motors do, try a higher voltage that stays within the recommended range. Bear in mind that if you change your batteries and any of the voltages, you'll need to check and reset your buck converter to ensure that you don't feed more than 5.1 V into your Raspberry Pi.

Alternatively, the motors themselves may just have a slow, geared RPM. If that's the case, while your robot may be slow, it will probably have a lot of torque, which is a fair trade-off.

Robot Not Following the Programmed Path

If your robot successfully executes the program and starts to move at a suitable speed, but doesn't follow the exact path you had planned, don't fret! Every motor is different and will need adjustments for the program to work the way you want. For example, 0.25 seconds may not be enough time for the motors to make your robot turn approximately 90 degrees. Edit the program and play around with the `sleep()` and `robot()` statements inside the `for` loop to adjust.

MAKING YOUR ROBOT REMOTE-CONTROLLED

Making a robot come to life and move is an exciting first step in robotics, and the natural next step is to make your robot remote-controlled. This means it will no longer be limited to a predefined path, so you'll be able to control it in real time!

The aim of this project is to program your robot so you can use a wireless controller to guide it. You'll be able to instantaneously change your robot's movements without going back into your code.

The Wiimote Wireless Controller

In order to control your robot with a wireless controller, first you'll need one! The perfect remote for our robot is a Nintendo Wii remote, also known as a *Wiimote*, like the one in Figure 4-3.

FIGURE 4-3

My much-loved Nintendo Wiimote

A Wiimote is a pretty nifty little Bluetooth controller with a set of buttons and some sensors that are able to detect movement. The Wiimote was originally created for the Nintendo Wii games console, but fortunately there's an open source Python library, called cwiid, that allows Linux computers, like your Raspberry Pi, to connect and communicate with Wiimotes. We'll use cwiid to manipulate the data from a Wiimote to control your robot's motors.

If you don't have a Wiimote already, you'll need to get your hands on one. These are widely available online, both new and used. I recommend picking up a cheap used one on a site like eBay or from a secondhand shop—mine cost me less than $15.

WARNING

To guarantee compatibility with your Raspberry Pi, make sure that your Wiimote is a Nintendo-branded official model. Over the years a considerable number of third-party Wiimotes have become available to buy. Though usually cheaper than an official Wiimote, these aren't guaranteed to work with the cwiid *library.*

You'll use Bluetooth to pair your Wiimote with the Raspberry Pi on your robot. *Bluetooth* is a wireless radio technology that many modern devices, like smartphones, use to communicate and transfer data over short distances. The latest Raspberry Pi models, like the Pi Zero W and Raspberry Pi 3 Model B+, come with Bluetooth capabilities built in. All models prior to the Raspberry Pi 3 Model B, like the original Raspberry Pi and Pi 2, do *not*, and consequently you'll need to get a Bluetooth USB adapter (or *dongle*), like the one pictured in Figure 4-4, to connect to a Wiimote.

FIGURE 4-4

A $3 Raspberry Pi–
compatible
Bluetooth dongle

These are available for less than $5 online; just search for "Raspberry Pi compatible Bluetooth dongle." Before you proceed, make sure you have plugged the dongle into one of the USB ports of your Pi.

Installing and Enabling Bluetooth

Before you start to write the next Python script, you'll need to make sure that Bluetooth is installed on your Pi and that the `cwiid` library is set up. Power your Raspberry Pi from a wall outlet and then, from the terminal, run this command:

```
pi@raspberrypi:~ $ sudo apt-get update
```

And then run this one:

```
pi@raspberrypi:~ $ sudo apt-get install bluetooth
```

If you have Bluetooth installed already, you should see a dialogue that states `bluetooth is already the newest version`. If you don't get this message, go through the Bluetooth installation process.

Next, you'll need to download and install the `cwiid` library for Python 3. We'll grab this code from *GitHub*, a website where programmers and developers share their software.

Run the following command in the home folder of your Pi:

```
pi@raspberrypi:~ $ git clone https://github.com/azzra/
python3-wiimote
```

You should now have the source code of the `cwiid` library downloaded to your Raspberry Pi, stored in a new folder called *python3-wiimote*. Before we can get to our next Python program, the source code must first be *compiled*, a process that makes and readies software for use on a device.

You also need to install four other software packages before you can proceed. Enter the following command to install all four at once:

```
pi@raspberrypi:~ $ sudo apt-get install bison flex automake
libbluetooth-dev
```

If you're prompted to agree to continue, press Y (which is the default). Once this command has finished executing, change into the newly downloaded directory containing your Wiimote source code:

```
pi@raspberrypi:~ $ cd python3-wiimote
```

Next, you must prepare to compile the library by entering each of the following commands, one after the other. This is all part of the compilation process—you don't have to worry about the specifics of each command! The first two commands won't output anything, but the rest of them will. I'll show the start of each output here:

```
pi@raspberrypi:~/python3-wiimote $ aclocal
```

```
pi@raspberrypi:~/python3-wiimote $ autoconf
```

```
pi@raspberrypi:~/python3-wiimote $ ./configure
checking for gcc... gcc
checking whether the C compiler works... yes
checking for C compiler default output file name... a.out
checking for suffix of executables...
--snip--
```

```
pi@raspberrypi:~/python3-wiimote $ make
make  -C libcwiid
make[1]: Entering directory '/home/pi/python3-wiimote/libcwiid'
--snip--
```

And then finally, to install the cwiid library, enter:

```
pi@raspberrypi:~/python3-wiimote $ sudo make install
make install -C libcwiid
make[1]: Entering directory '/home/pi/python3-wiimote/libcwiid'
install -D cwiid.h /usr/local/include/cwiid.h
--snip--
```

NOTE

If you have trouble with the Python 3 cwiid installation, check out the book's website to see whether the process has been updated: https://nostarch.com/ raspirobots/.

After that, cwiid should work in Python 3! Now you can navigate out of the *python3-wiimote* directory and back to where you have all of your other code.

Programming Remote Control Functionality

Now create and open a new Python program to store the Wiimote code. I have called mine *remote_control.py*:

```
pi@raspberrypi:~/robot $ nano remote_control.py
```

In general, before you start to code, it is important to first plan what exactly you want to do. In our case, we want to think about how we want the Wiimote to control the robot exactly. Let's make a plan.

The Wiimote has 11 digital buttons, which is more than we'll need for this simple project. Interestingly for us, 4 of those buttons belong to the D-pad—the four-way directional control buttons at the top of your Wiimote, shown in Figure 4-5.

FIGURE 4-5

The four-way D-pad of a Wiimote

That's perfect for our purposes: we can use up to make the robot go forward, right to make the robot go right, down to make the robot go backward, and left to make the robot go left. This is very similar to the program we wrote earlier, except that now we read our inputs from the Wiimote rather than them being programmed in.

We also need something to make the robot stop. The "B" trigger button on the underside of the Wiimote would be well suited to this. Let's write some code in Nano that executes the plan we've made; see Listing 4-2. I have saved this program as *remote_control.py*.

LISTING 4-2

Programming your robot to respond to the D-pad of your Wiimote

```
import gpiozero
import cwiid

❶ robot = gpiozero.Robot(left=(17,18), right=(27,22))

print("Press and hold the 1+2 buttons on your Wiimote simultaneously")
❷ wii = cwiid.Wiimote()
print("Connection established")
❸ wii.rpt_mode = cwiid.RPT_BTN

while True:
    ❹ buttons = wii.state["buttons"]

    ❺ if (buttons & cwiid.BTN_LEFT):
        robot.left()
    if (buttons & cwiid.BTN_RIGHT):
        robot.right()
    if (buttons & cwiid.BTN_UP):
        robot.forward()
    if (buttons & cwiid.BTN_DOWN):
        robot.backward()
    if (buttons & cwiid.BTN_B):
        robot.stop()
```

As before, you start by importing gpiozero as well as the new cwiid library. A Robot object is then set up ❶.

In the next section of code ❷, you set up the Wiimote. As with the Robot object, we assign the Wiimote object to a variable called wii. When this code runs and execution reaches this line, there will be a pairing handshake between the Raspberry Pi and Wiimote. The user will need to *press and hold* buttons 1 and 2 on the Wiimote at the same time to put the Wiimote in a Bluetooth-discoverable mode. We add a print() statement here to tell the user when to press the buttons.

If the pairing is successful, the code prints a positive message for the user. We then turn on the Wiimote's reporting mode ❸, which permits Python to read the values of the different buttons and functions.

After this, we use an infinite `while` loop to tell the robot what to do when each button is pressed. First, the loop reads the current status of the Wiimote ❹, meaning it checks what buttons have been pressed. This information is then stored in a variable called `buttons`.

Finally, we start the last chunk of the program ❺: a variety of `if` statements and conditions that allocate an action to each button. To look at one example, the first `if` statement ensures that if the left button of the D-pad has been pressed, the robot is instructed to turn left. Over the next lines, the same sort of logic is applied: if the right button of the D-pad has been pressed, the robot is instructed to turn right, and so on.

As usual, once you have finished writing your program, exit Nano and save your work.

Running Your Program: Remote-Control Your Robot

Place your robot on a large surface and have your Wiimote handy. If your Pi requires a Bluetooth dongle, don't forget to plug it into one of the USB ports. To run your program, use an SSH terminal to enter:

```
pi@raspberrypi:~/robot $ python3 remote_control.py
```

Soon after program execution, a prompt will appear in the terminal asking you to press and hold the 1 and 2 buttons on your Wiimote simultaneously. You should hold these buttons until you get a success message, which can take up to 10 seconds. The Bluetooth handshake process can be fussy, so try to press them as soon as the program instructs you to do so.

If the pairing was successful, another message stating `Connection established` will appear. Alternatively, if the pairing was unsuccessful, an error message saying that `No Wiimotes were found` will be displayed, and your program will crash. If this is the case, and you are using an official Nintendo-branded Wiimote, then you most likely were not fast enough pressing the 1 and 2 buttons! Rerun the program with the same command and try again.

With your Wiimote now successfully connected, you should be able to make your robot dash around in any direction you want at the touch of a button! Remember that you can stop both motors at any point by pressing B on the underside of your Wiimote. As usual, you can kill the program by pressing CTRL-C.

VARYING THE MOTOR SPEED

Up until now your robot has been able to go at two speeds: 0 mph, or top speed! You might have noticed that this isn't the most convenient. Traveling at full speed makes precise maneuvers almost impossible, and you probably crashed into things a few times. Fortunately, it doesn't always have to be this way. Let's give your robot some control over its speed.

In this project, we'll build upon the previous example and create a remote control robot with variable motor speed. To do this I'll introduce a technique called *pulse-width modulation (PWM)*, and I'll explain how to use it inside the Python GPIO Zero library. We'll also put a special sensor called an *accelerometer* in your Wiimote to good use to create a much improved version of the remote control program!

Understanding How PWM Works

The Raspberry Pi is capable of providing *digital* outputs but not *analog* outputs. A digital signal can be either on or off, and nothing in between. An analog output, in contrast, is one that can be set at no voltage, full voltage, or anything in between. On the Raspberry Pi, at any given time a GPIO pin is either on or off, which is no voltage or full voltage. By this logic, motors connected to a Pi's GPIO can only either stop moving or go full speed.

That means that it is impossible to set a Pi's GPIO pin to "half voltage" for half the motor speed, for example. Fortunately, the PWM technique allows us to approximate this behavior.

To understand PWM, first take a look at the graph in Figure 4-6. It depicts the state of a digital output changing from low to high. This is what happens when you turn on one of your Pi's GPIO pins: it goes from 0 V to 3.3 V.

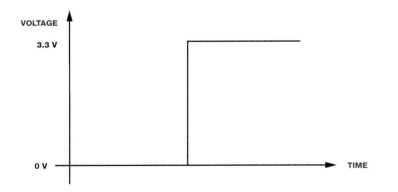

FIGURE 4-6

A state change from low (0 V) to high (3.3 V)

PWM works by turning a GPIO pin on and off so quickly that the device (in our case, a motor) "notices" only the *average* voltage at any given time. This means that the state is somewhere in between 0 V and 3.3 V. This average voltage depends on the *duty cycle*, which is simply the amount of time the signal is on, versus the amount of time a signal is off in a given period. It is given as a percentage: 25 percent means the signal was high for 25 percent of the time and low for 75 percent of the time; 50 percent means the signal was high for 50 percent of the time and low for the other 50 percent, and so on.

The duty cycle affects the output voltage proportionally, as shown in Figure 4-7. For example, for the Raspberry Pi, pulse-width modulating a GPIO pin at a 50 percent duty cycle would give a voltage of 50 percent: 3.3 V / 2 = 1.65 V.

FIGURE 4-7

Two different PWM voltage traces: a duty cycle of 25 percent (top) and a duty cycle of 50 percent (bottom)

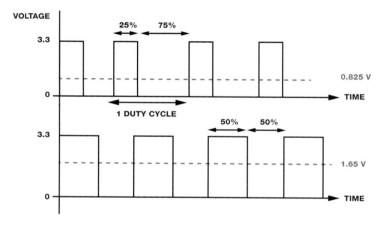

While PWM is not a perfect approximation of an analog signal, for most cases it works well, especially at this level. Digitally encoding analog signal levels will allow you to control the exact speed of your robot's movement.

The GPIO Zero Python library authors have made it easy to vary motor speed using PWM, so you don't need to know the exact mechanics behind it. All you need to do is provide a value between 0 and 1 in the parentheses of each motor command to represent a value between 0 percent and 100 percent, as follows:

```
robot.forward(0.25)
time.sleep(1)
robot.left(0.5)
```

```
time.sleep(1)
robot.backward()
time.sleep(1)
```

This program would command your robot to move forward for
1 second at 25 percent of its full speed, turn left at 50 percent of its full
speed for another second, and then go backward at full speed for a final
second. If you don't provide a value, Python assumes that the robot
should move at full speed, just the same as if you were to enter a 1.

Understanding the Accelerometer

Before we improve upon the remote control program in the previous
project, let's learn about the accelerometer in your Wiimote and how
we can use it.

Previously, you used the D-pad of the Wiimote to provide control.
These four buttons are digital and can only detect being pressed on
or off. This isn't ideal for controlling both speed and direction at once.

Inside each Wiimote, however, there is a sensor called an *accelerometer* that can detect and measure the acceleration the Wiimote is
undergoing at any point. This means that moving a Wiimote in the air
provides sensory data in all three axes: in all three axes: *x*, *y*, and *z*. In
this way, the accelerometer can track the direction of movement, and
the speed of that direction. See Figure 4-8 for a diagram.

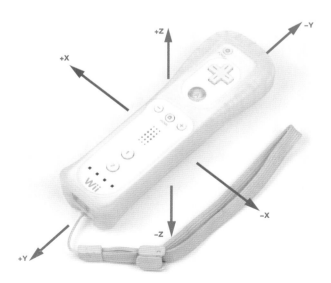

NOTE

*If your robot has been
zipping around too fast in
the previous examples, feel
free to go back and adjust
the speed in the last two
projects using this method!*

FIGURE 4-8

The axes of motion the
Wiimote's accelerometer
can detect

This kind of analog data is ideal for a variable-motor-speed
remote control program. For example, the more you pitch the
Wiimote in the *x* direction, the faster your robot could move forward.

Looking at the Data

Before we rework the robot's program, it would be incredibly helpful to see the raw data that the accelerometer from the Wiimote outputs. Once we have an idea of what that output looks like, we can think about how to manipulate that data to correspond to the robot's movement.

Power the Pi on your robot from a wall outlet, open a new file in Nano and call it *accel_test.py*, and then enter the code in Listing 4-3—this script uses the cwiid library too, so if you haven't installed that, see the instructions in "Installing and Enabling Bluetooth" on page 88.

LISTING 4-3

The code to print raw accelerometer data

```
import cwiid
import time

❶ print("Press and hold the 1+2 buttons on your Wiimote simultaneously")
wii = cwiid.Wiimote()
print("Connection established")
❷ wii.rpt_mode = cwiid.RPT_BTN | cwiid.RPT_ACC

while True:
  ❸ print(wii.state['acc'])
    time.sleep(0.01)
```

This simple program prints the Wiimote's accelerometer data to the terminal every 0.01 seconds.

The print() statement denotes the start of the Wiimote setup ❶. The three following lines are the same as in the prior project, with the exception of the final line in that code block ❷, with which we're not just turning on a Wiimote's reporting mode like before, but also permitting Python to read values from both the buttons *and* the accelerometer. If you haven't come across it before, the keyboard character in the middle of this line is called a *vertical bar* or a *pipe*. It is likely to be located on the same key as the backslash on your keyboard.

An infinite while loop prints the status of the accelerometer ❸. The next line waits for 0.01 seconds between each *iteration* of the while loop so that the outputted data is more manageable. In programming, each time a loop goes round and executes again is called an iteration.

You can run this program with the command:

```
pi@raspberrypi:~/robot $ python3 accel_test.py
```

After you pair your Wiimote, accelerometer data should start printing to the terminal. The following output is some of the data that I saw in my terminal:

```
(147, 123, 136)
(151, 116, 136)
(130, 113, 140)
(130, 113, 140)
(130, 113, 140)
```

Each line of data is delivered as three values in parentheses, representing the x-, y-, and z-axes, respectively, which change as you move the Wiimote in the different axes. Experiment with different movements and watch as the figures go up and down. Exit the program by pressing CTRL-C.

With this raw data, we can put some thought into the next part of the programming process, namely answering the question: How can you translate those three figures into instructions for your robot? The best way to approach this problem is logically and in small steps.

Figuring Out the Remote Movement Control

First, consider the movement of your two-wheeled robot. Because it moves around only on the floor, and doesn't fly up and down, its movement can be expressed in two dimensions: *x* and *y*, as shown in Figure 4-9. We can disregard the z-axis data, which simplifies the problem substantially.

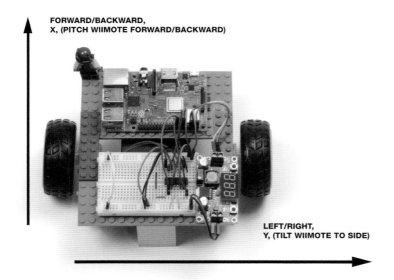

FORWARD/BACKWARD,
X, (PITCH WIIMOTE FORWARD/BACKWARD)

LEFT/RIGHT,
Y, (TILT WIIMOTE TO SIDE)

FIGURE 4-9

Only two axes of control are required for your two-wheeled robot.

Second, consider how you wish to hold the Wiimote when controlling your robot. I have decided to hold it horizontally, with the 1 and 2 buttons close to my right hand, as shown in Figure 4-10. This is the most common orientation for traditional Wii-based racing games and is ideal for controlling your robot.

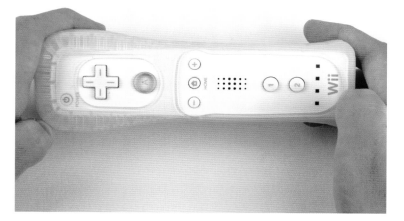

When you're holding the Wiimote in this way, pitching it backward and forward controls the x values. Tilting it side-to-side controls the y values.

When you printed your accelerometer data, you may have noticed that the outputted numbers tended to be between 95 and 145. You can run the test program again to observe this. This is because the lowest x value is 95, when the Wiimote is pitched all the way back. This highest value is 145, when it's pitched entirely forward.

For the y-axes, left to right, the lowest value is 95 and the highest is 145. The difference between 145 and 95 is 50, and this gives us the usable range of data in each axis. See Figure 4-11 for an illustration of how the Wiimote's values change.

So far in this chapter, you've controlled your robot's movement by instructing it to go forward, backward, left, or right at full speed. We want to change this to vary the speed according to the accelerometer. Luckily, the Robot class from the GPIO Zero Python library has another way of turning the motors on and setting their speed that suits our needs.

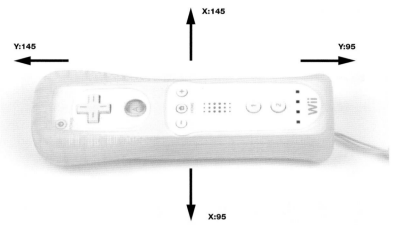

FIGURE 4-11

The Wiimote's extreme accelerometer values

The Robot class has an *attribute*—a variable that is part of a class—called value. At any given time, value represents the motion of the robot's motors as a pair of numeric values between –1 and 1. The first value in the pair is for the left motor's speed, while the second value is for the right motor's speed. For example, (–1, –1) represents full speed backward, whereas (0.5, 0.5) represents half speed forward. A value of (1, –1) would represent turning full speed right. By setting the value attribute, you can manipulate your robot in any direction you wish. This is going to come in super-handy in the upcoming program!

Programming Your Robot for Variable Speed

Now that we've broken down this problem and found a neat and efficient final approach to the program, we can start coding! Use Nano to create a new program called *remote_control_accel.py* and input the code shown in Listing 4-4.

```
import gpiozero
import cwiid

robot = gpiozero.Robot(left=(17,18), right=(27,22))

print("Press and hold the 1+2 buttons on your Wiimote simultaneously")
wii = cwiid.Wiimote()
print("Connection established")
wii.rpt_mode = cwiid.RPT_BTN | cwiid.RPT_ACC
```

```
while True:
❶   x = (wii.state["acc"][cwiid.X] - 95) - 25
    y = (wii.state["acc"][cwiid.Y] - 95) - 25

❷   if x < -25:
            x = -25
    if y < -25:
            y = -25
    if x > 25:
            x = 25
    if y > 25:
            y = 25

❸   forward_value = (float(x)/50)*2
    turn_value = (float(y)/50)*2

❹   if (turn_value < 0.3) and (turn_value > -0.3):
            robot.value = (forward_value, forward_value)
    else:
            robot.value = (-turn_value, turn_value)
```

NOTE

Python (and many other programming languages) can deal with numbers in different ways. The main two number types in Python are called integers and floats. Integers are whole numbers that have no decimal point. Floats (floating-point real values) have decimal points and can represent both the integer and fractional part of a number. For example, 8 is an integer, whereas 8.12383 or 8.0 is a float. In the remote_control_accel.py program, we need to use floats, as the movement of your robot will be governed by two numbers in between –1 and 1.

The program shares the same Wiimote setup process as the accelerometer test program. Then we set up a while loop to keep running our code. The first statement ❶ reads the x value from the accelerometer and then stores it in a variable called x. Within the variable, the value undergoes two arithmetic operations. First, 95 is subtracted; this limits the data to a value between 0 and 50, rather than between 95 and 145, so that it fits within the usable range we discovered earlier. Second, we subtract a further 25. This ensures the range of data will be between –25 and +25. Exactly the same process then happens for the y value, and the result is then stored in a variable called y.

We need to do this because the value attribute of the Robot class accepts negative values for backward movement and positive values for forward movement. This manipulation balances the accelerometer data on either side of 0, making it clear which values are for reverse and which are for forward movement.

The four if statements ❷ eliminate the chance for errors later in the program. In the unlikely event that the Wiimote's accelerometer outputs data that is not within the –25 to +25 range, the if statements catch this occurrence and then round up or down to the relevant extremity.

Next, the final x-axis value for the robot is determined and stored in a variable called forward_value ❸. This calculation divides the x variable value by 50, providing a new proportional number between −0.5 and 0.5. This result is then multiplied by 2 to get a value between −1 and 1. The same process is repeated to get the final y-axis value, which is then stored in a similar variable called `turn_value`.

The line at ❹ starts an `if/else` clause. If the `turn_value` is less than 0.3 or greater than −0.3, `robot.value` is set to be the `forward_value`. So, if the Wiimote is tilted by *less than 30 percent* to either side, the program will assume that you want the robot to move forward/backward. This means that your robot won't turn in the wrong direction at the slightest tilt of your Wiimote. The forward/backward speed of your robot is then set according to the pitch of your Wiimote. For example, if your Wiimote is pitched all the way forward, it will set `robot.value` to (1, 1) and your robot will accelerate forward.

Alternatively, if the Wiimote is tilted by more than 30 percent to either side, the program will assume that you want the robot to turn left or right on the spot. The program then turns the robot based on the angle of your Wiimote tilt. For example, if you have the Wiimote tilted all the way to the right, your robot will spin very quickly to the right; but if you have it tilted only slightly to the right, the robot will turn more slowly and in a more controlled manner.

As usual, after you have finished your program, exit Nano and save your work.

Running Your Program: Remote-Control Your Robot with PWM

Disconnect your robot from your wall outlet, and ensure that it is powered by its batteries. Then place it on a large surface and have your Wiimote in hand and in a horizontal orientation. To run your program, enter:

```
pi@raspberrypi:~/robot $ python3 remote_control_accel.py
```

After you have gone through the familiar Bluetooth handshake process, your robot should come to life and start to move as you change the orientation of your Wiimote. Experiment with driving it around at different speeds and practice maneuvering!

Challenge Yourself: Refine your Remote-Controlled Robot

When you have a feel for the behavior of your remote-controlled robot, take another look at the code and refine it as you see fit. For example, you could try to make the steering more sensitive, limit the speed of your robot, or even make your robot move in a predefined pattern when you press a button. The possibilities are endless!

SUMMARY

This chapter has taken you from having a robot-shaped paperweight to a fully functional Wiimote-controlled little machine! We have covered a wide range of concepts from H-bridges to PWM to accelerometers. Over the process you have written three programs, each more advanced than the last.

In the next chapter, I'll guide you through making your robot a little bit more intelligent so that it can automatically avoid obstacles!

5
AVOIDING OBSTACLES

NOW YOU CAN CONTROL HOW YOUR ROBOT MOVES, AND THAT'S PRETTY COOL! BUT WOULDN'T IT BE COOLER STILL TO GET YOUR ROBOT TO CONTROL ITSELF?

You may have noticed that your little two-wheeler is quite vulnerable to all kinds of hazards while it's running around the floor. Crashing into walls and other objects can be very annoying and even dangerous to your hardware. In this chapter, I'll show you how to enable your robot to autonomously detect and avoid obstacles. We'll cover the theory behind obstacle detection and how to use the sensor you'll need.

OBSTACLE DETECTION

In order for your robot to avoid obstacles, it will first need to be able to sense them. In electronics we use specialized sensors for this purpose. There's a variety of ways to implement obstacle detection using sensors. At the hobbyist level, there are two main approaches: digital detection and analog detection. Digital detection is excellent at sensing obstacles within a certain range, but it can't determine the distance to that obstacle. Analog detection, on the other hand, can do both, so that's what we'll use here to make our robot extra intelligent.

Using Ultrasonic Sensors for Analog Object Detection

The HC-SR04 ultrasonic sensor (seen in Figure 5-1) uses ultrasonic sound to determine the distance between the sensor and an object. The sensor works in much the same way that bat and dolphin sonar works in the natural world, and submarine sonar works in the not-so-natural world.

FIGURE 5-1

The HC-SR04 ultrasonic distance sensor

Sound can be modeled as a wave with varying wavelength. Only a small range of the sound spectrum is audible to the human ear. Any sound waves with a frequency above this range (20 kHz+) are *ultrasound* waves. Ultrasonic distance sensors are designed to sense object proximity using ultrasound reflection. A sonar system like this sends out waves that bounce off obstacles. A receiver then detects the returning sound waves. Ultrasound is accurate within short distances (around a few meters) and is inaudible to humans.

Understanding How the HC-SR04 Works

A basic ultrasonic distance sensor like the HC-SR04 is made up of a transmitter, a receiver, and some circuitry. The transmitter and the receiver are the speaker-like protrusions resembling eyes in Figure 5-1.

To determine a distance, the transmitter emits a high-frequency ultrasonic sound. This sound will bounce off any nearby solid objects and be reflected. The "bounce-back" is detected and picked up by the receiver on the HC-SR04.

Sound travels through air at a constant speed. At room temperature (20° C/68 °F) a sound wave will travel at approximately 343 m/s (meters per second). While this is fast, it's not instantaneous, which means there's a small time difference between when the sound is emitted and when the bounce-back is received. Therefore, we can measure the distance by timing how long the signal takes to bounce back to the sensor.

The relationship between speed, distance, and time can be summarized as follows:

$$\text{speed } (m/s) = \frac{\text{distance } (m)}{\text{time taken } (s)}$$

The speed of an object in meters per second is equal to the distance that object has traveled in meters divided by the time it has taken to move that far in seconds. We'll use this equation to figure out the distance. We know the speed of sound is constant at 343 m/s, and we can measure how long the sound wave takes to bounce off an object, which is the time. If you rearrange the equation to solve for distance, you get:

$$\text{distance } (m) = \text{speed } (m/s) \times \text{time } (s)$$

This isn't quite the full story, however. The ultrasonic pulse is emitted, bounces off an object, and is received by the HC-SR04, as shown in Figure 5-2, meaning the sound wave is actually traveling *double* the distance from the sensor to the object.

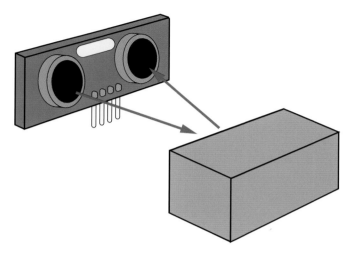

That means we need to divide the time recorded by the ultrasonic distance sensor in half, which results in the following equation:

$$\text{distance } (m) = \text{speed } (m/s) \times \frac{\text{time } (s)}{2}$$

We have our method, so let's try it out.

MEASURING A SHORT DISTANCE

Now that you understand the mathematics and theory behind ultrasonic distance measurement, it's time to put everything into practice and measure some distances!

The Parts List

In addition to the breadboard you already have on your Raspberry Pi robot, you'll need the following components:

- An HC-SR04 ultrasonic sensor
- A 1 kΩ resistor
- A 2 kΩ resistor
- Jumper wires

Inexpensive HC-SR04 sensors are widely available through the usual online retailers. Just search "HC-SR04;" you shouldn't have to spend more than a few dollars on one.

Any digital system has two logic states: low voltage (0) and high voltage (1). This was first introduced in Chapter 4 when explaining PWM. Usually, the low voltage is just ground, 0 V; however, the high voltage can change from system to system. This means that some systems require a 5 V signal to trigger a high voltage, while others may require only 3.3 V, for example. This just so happens to be the situation we are in! The HC-SR04 requires 5 V, whereas your Raspberry Pi operates on 3.3 V logic. Notice in Figure 5-3 that there are four pins on the ultrasonic sensor: *Vcc* for power, *Trig* for the trigger pulse, *Echo* for the echo pulse, and finally *Gnd* for ground.

FIGURE 5-3

A close-up of the pins on an HC-SR04 module

You'll need to power the module using a 5 V source on the Vcc pin. When the HC-SR04 receives the bounce-back pulse, the Echo pin is set to a 5 V logic high, but if we were to connect this directly to the Raspberry Pi, the high voltage would cause serious damage. To avoid this, you'll need to lower the sensor output voltage to something your Raspberry Pi can handle—that's where the 1 kΩ and 2 kΩ resistors come in handy. We're going to use them to build a *voltage divider*.

Reducing Voltage with Voltage Dividers

A *voltage divider* is a simple circuit that turns a larger voltage into a smaller one. The voltage divider takes an input voltage and uses two resistors in series to reduce and output that voltage. You use different values of resistors to make the output voltage a certain fraction of the input voltage.

The voltage divider circuit is shown in Figure 5-4. Notice that the output voltage is drawn from between the two resistors.

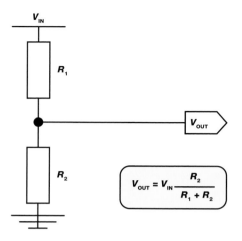

FIGURE 5-4

A voltage divider circuit

As with most electronics, we use an equation to mathematically relate the two resistors (R_1 and R_2) to the input and output voltages (V_{in} and V_{out}). Physicists don't mince words, so the equation is simply called the *voltage divider equation*:

$$V_{out} = V_{in} \times \frac{R_2}{(R_1 + R_2)}$$

Using this equation, you can work out the exact R_1 and R_2 resistor values you need to create the desired output voltage. The size of R_1 and R_2 is actually irrelevant; what matters instead is the *ratio* of R_1 and R_2. For example, if R_1 and R_2 are equal, then the output voltage will be half the input voltage.

WARNING

No matter what resistor values you end up with, it's always better to have an output voltage that is lower than the target, rather than higher. This is because a lower voltage will not harm your Raspberry Pi or other electronics, whereas a voltage that is too high, even by a small margin, may do so.

Let's use this equation to work out what resistor values we need for the distance sensor on our robot. You know that the input voltage is 5 V, and the desired output voltage is 3.3 V, so we have two unknowns in the equation: R_1 and R_2. You can pick a common resistor value for one and that gives us just one unknown, which is much easier to work out. Let's choose 1 kΩ for R_1. By rearranging the equation to solve for R_2, you get:

$$R_2 = \frac{(V_{out} \times R_1)}{V_{in} - V_{out}}$$

If you plug in the numbers that you know, R_2 works out like this:

$$R_2 = \frac{(3.3 \times 1,000)}{(5 - 3.3)}$$

$$R_2 = 1941.176471 \ \Omega$$

Finding a resistor with the exact value 1941.176471 Ω is going to be very tricky! So instead we'll just pick the nearest common resistor value. In our case, a 2 kΩ or 2.2 kΩ resistor will suffice. If in doubt, try to find a resistor that's nearest, but slightly *lower* than, the value you need. You can always put your two resistor values back into the equation to work out the output voltage with that pair if you want to double-check.

Wiring Up Your HC-SR04

Now that you have all your necessary components, you can wire up your distance sensor. As always, ensure your Pi and robot's power are disconnected before you start fiddling around with the wiring and connecting new things.

Rather than plug the HC-SR04 module directly into a breadboard, we'll connect it using long jumper wires so you can position the distance sensor anywhere on your robot. Using your build from the previous chapter that has the motors wired up, follow these instructions, but don't attach the sensor to the robot chassis just yet:

1. Use a long jumper wire to connect the Vcc pin of your HC-SR04 to the Pi's +5 V power rail on your breadboard.

2. Use another long jumper wire to connect the Gnd pin of your HC-SR04 to the ground rail on your breadboard. At this point your wiring should look like Figure 5-5.

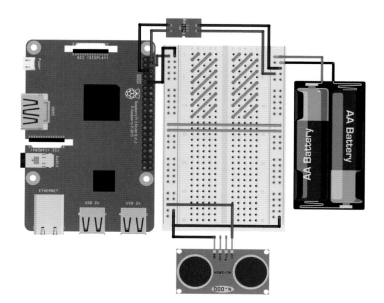

FIGURE 5-5

HC-SR04 module connected to +5 V and ground. The motor controller chip and motors are omitted from the diagram, but you should keep them connected to your circuit. The orange stripes represent the area the L293D chip and its wiring takes up.

3. Next, use a wire to connect the Trig pin of your HC-SR04 directly to physical pin 16 on your Raspberry Pi. Pin 16 on the Pi is also called BCM 23.

4. Then attach a jumper wire from the Echo pin of your sensor to a new row on your breadboard. Connect a 1 kΩ resistor by putting one leg of the resistor into the same row as the Echo pin and the other leg of the resistor into a different, unused row of your breadboard. At this point your wiring should look like Figure 5-6.

5. Wire up physical pin 18 (BCM 24) of your Raspberry Pi to the row with the other leg of the 1 kΩ resistor you just connected.

6. Finally, place one leg of your 2 kΩ/2.2 kΩ resistor in the row with the 1 kΩ resistor and jumper wire to your Pi's BCM 24 pin, and place the other leg of this resistor in a ground rail. The complete circuit on your breadboard should look something like Figure 5-7.

Figure 5-8 shows a circuit diagram of the final outcome.

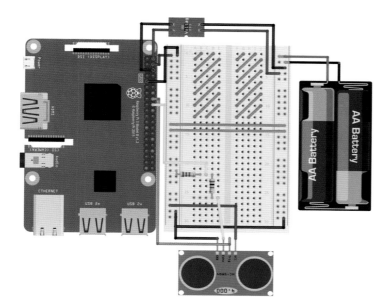

FIGURE 5-7

The completed breadboard diagram with the HC-SR04 and voltage divider in place

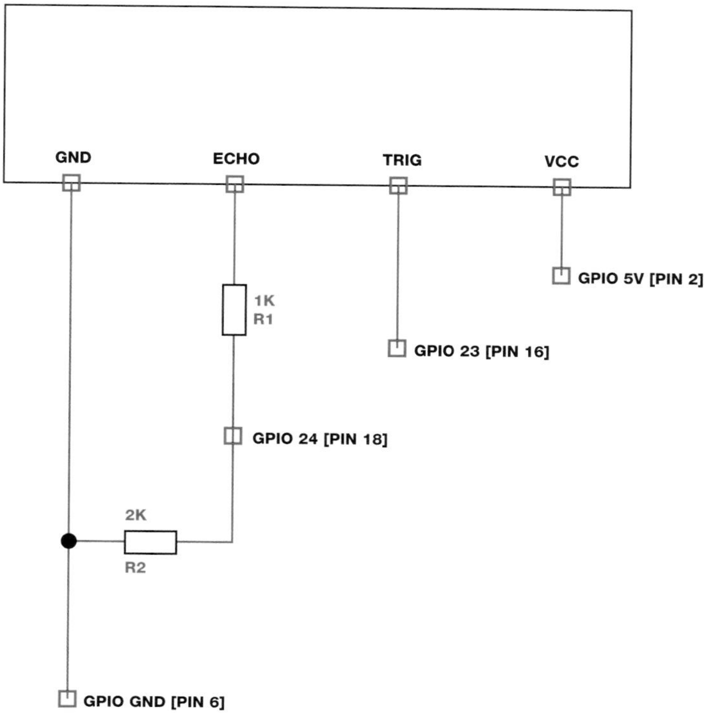

FIGURE 5-8

A circuit diagram of the HC-SR04 connected to a Raspberry Pi

GND   ECHO   TRIG   VCC

                                         GPIO 5V [PIN 2]

                  1K
                  R1

                                GPIO 23 [PIN 16]

                  GPIO 24 [PIN 18]

         2K
         R2

GPIO GND [PIN 6]

Programming Your Raspberry Pi to Read Distance

With your ultrasonic distance sensor now wired up, it's time to delve into some more Python code to use the sensor. Boot up your Raspberry Pi from a wall outlet, log in, and locate the *robot* folder you're saving your programs in. Create a new program called *distance_test.py* with this command:

```
pi@raspberrypi:~/robot $ nano distance_test.py
```

In this project, we'll still use the GPIO Zero Python library, but we'll move away from the built-in functions and objects. Instead, I'll show you how to program and operate the HC-SR04 entirely from scratch!

The code in Listing 5-1 will send out a signal, also known as a *ping*, and then read and print out the distance of the first object the signal meets. Save this code into the *distance_test.py* you have already open. Try to use your programming skills to read through and decipher it before moving on to my explanation!

LISTING 5-1

Program to measure a single distance

```python
import gpiozero
import time

❶ TRIG = 23
  ECHO = 24

❷ trigger = gpiozero.OutputDevice(TRIG)
  echo = gpiozero.DigitalInputDevice(ECHO)

❸ trigger.on()
  time.sleep(0.00001)
  trigger.off()

❹ while echo.is_active == False:
      pulse_start = time.time()

❺ while echo.is_active == True:
      pulse_end = time.time()

❻ pulse_duration = pulse_end - pulse_start

❼ distance = 34300 * (pulse_duration/2)

  round_distance = round(distance,1)

  print("Distance: ", round_distance)
```

As usual, we begin by importing the `gpiozero` and `time` libraries for use throughout the code.

At ❶ and on the following line, we create two variables, TRIG and ECHO, which simply store the pin number that the Trig and Echo pins are connected to, respectively. We capitalize these variables to indicate that they are *constants*—variables whose values we want to keep the same for the duration of the program.

Capitalizing constants is a programming convention that tells you, and anyone else reading the code, that these values are not changed throughout the execution of the program. It's worth stressing that because this is a convention, it's merely a practice that most programmers do, not one enforced by Python. The code would work equally well if these were lowercase, or even a mixture of cases.

The following two lines ❷ set up the GPIO pins connected to the Trig and Echo pins of the HC-SR04. We set up the `trigger` variable to be an output, since it's sending out a ping, and `echo` as an input, since it's receiving a ping.

To trigger the ping, the HC-SR04 sensor needs a quick 10 µs (1 µs is a millionth of a second: 0.000001 s) pulse that takes place in the next chunk of code ❸.

When the ping has been sent out by the transmitter, the program must wait for the ping to clear the receiver before it starts listening for the echo. This is because the transmitter and the receiver on the HC-SR04 are close together, so in the microseconds after transmission, the sensor can hear the outgoing pulse. We don't want to record the outgoing pulse accidentally, so we tell the program to wait until it hears the echo *and not* the original ping. You can think of this as if you and a friend were standing next to each other in a large room and you wanted to listen for an echo. If you were to shout "Hello!" your friend would hear you *before* you heard the echo from the room. This sort of effect is what we must avoid when using the HC-SR04.

The next section of the code is responsible for making sure we pick up the echo. The `while` loop ❹ with the condition `echo.is_active == False` repeats until the outgoing pulse is no longer heard by the sensor. The program then stores the exact time the pulse clears the receiver in a new variable called `pulse_start`.

With the outgoing pulse now out of the way, the `while` loop with the condition `echo.is_active == True` ❺ catches the echo when it returns to the sensor. A second variable called `pulse_end` is created and is used to record the exact time of the bounce-back pulse.

Then we simply subtract the time the ping was sent out from the time it was received to work out how long it took to return. We store the result in a variable called `pulse_duration` ❻.

Arguably the most important part of this program is at ❼ where we work out the distance from the time it has taken for the ping to return. We apply the earlier equation to the values we've collected in the program:

$$\text{distance } (m) = \text{speed } (m/s) \times \frac{\text{time } (s)}{2}$$

Rather than use the 343 m/s figure that is the speed of sound, we multiply it by 100 to give us a distance value in centimeters, which is much more relevant to the sorts of distances your robot will be dealing with.

Finally, in the last few lines of the program, we round the distance value to one decimal place and then output it to the terminal.

Running Your Program: Measure a Short Distance

Now that you have the code finished, it's time to test out the ultra-sonic distance sensor and its accuracy.

Place your HC-SR04 parallel to a surface, like a table. Then place a fairly small solid object in front of the sensor and measure the distance with a ruler. In Figure 5-9, I'm using an upright box. My box is roughly 20 cm away from my HC-SR04.

FIGURE 5-9

My HC-SR04 test setup

As ever, to run your program, enter:

```
pi@raspberrypi:~/robot $ python3 distance_test.py
```

After a short period, a single distance reading should print to your terminal and the program will end. For me this looks like the following:

```
pi@raspberrypi:~/robot $ python3 distance_test.py
Distance: 20.1
```

Your HC-SR04 should have successfully measured the space between itself and the object! Mine was pretty accurate, but you shouldn't expect 100 percent accuracy with these readings.

If your reading was off by a large margin, try running the program again to see if that output was just an outlier. If you're still getting wrong readings, check the numbers and equations in your program: are the values correct and have you applied the math in the right way? If the program hangs (does nothing) and never finishes executing, check that your wiring is correct and refer to the instructions earlier in the chapter. Your program may also hang if it never receives an echo. This could be because the distance you're trying to measure is out of range. However, for indoor use you shouldn't have any problems.

Finally, if you're still having issues, consult the code and make sure it's exactly the same as the program in Listing 5-1. As usual, you can grab the exact code examples from *https://nostarch.com/raspirobots/*.

MAKE YOUR ROBOT AVOID OBSTACLES

Now that you have mastered measuring individual distances with an ultrasonic distance sensor, you can mount the sensor onto your robot and write a new program that will use the HC-SR04 to avoid obstacles.

By the end of this project, you'll have a fully autonomous obstacle-avoider! The aim here is to make sure your Raspberry Pi robot gets no closer than 15 cm to any object before it takes evasive action.

Mounting Your HC-SR04 Ultrasonic Sensor

The best place to mount your distance sensor is on the front of the robot in a location that is as central as possible. I recommend using sticky tack or double-sided tape to affix it. The module can sense distance only in a direct straight line, so don't mount it too high above the floor or your robot is likely to crash into low-lying obstacles.

I have mounted my HC-SR04 on the front stabilizer, as shown in Figure 5-10. The sensor is about a centimeter off the ground. Note that the orientation of your sensor also doesn't matter: mine is upside down!

FIGURE 5-10

My HC-SR04 mounted to the front of my robot

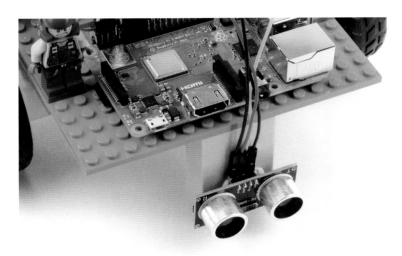

Programming Your Robot to Avoid Obstacles

To create the obstacle avoidance program, we'll borrow heavily from the prior section's code but set the distance sensor to constantly scan for upcoming obstacles.

In Listing 5-1, we worked out how to measure a single distance in 10 lines of code. In our next program, we'll need to repeat the code continuously to get a constantly refreshing distance from your robot to any upcoming obstacles. We could write out the code every time we need to use it, but that sounds time-consuming and dull, and we don't even know how many times we'd need to write it out. Instead, there is a way to package up code so you can use it whenever and wherever you need it. Packaging up code like this creates what is known as a *function*.

A Python function is a block of organized, reusable code that performs an action. Let's try it out: enter the code for the obstacle avoidance program in Listing 5-2 and save it as *obstacle_avoider.py* on your Raspberry Pi.

LISTING 5-2

Obstacle avoidance
program

```
   import gpiozero
   import time

❶ TRIG = 23
   ECHO = 24

   trigger = gpiozero.OutputDevice(TRIG)
   echo = gpiozero.DigitalInputDevice(ECHO)

❷ robot = gpiozero.Robot(left=(17,18), right=(27,22))

❸ def get_distance(trigger, echo):
❹     trigger.on()
       time.sleep(0.00001)
       trigger.off()

       while echo.is_active == False:
           pulse_start = time.time()

       while echo.is_active == True:
           pulse_end = time.time()

       pulse_duration = pulse_end - pulse_start

       distance = 34300 * (pulse_duration/2)

       round_distance = round(distance,1)

❺     return(round_distance)

   while True:
❻     dist = get_distance(trigger,echo)
❼     if dist <= 15:
           robot.right(0.3)
           time.sleep(0.25)
❽     else:
           robot.forward(0.3)
           time.sleep(0.1)
```

This program begins by importing the necessary libraries. Then, you identify ❶ and set up ❷ the Trig and Echo pins of the HC-SR04 like you did in Listing 5-1. This also initializes the robot for use.

At ❸, you meet your first Python function, which is organized into a code block. To start a function block, you use the keyword def. This is short for *define*, as you are defining what the block of code should do.

After def, you enter the name of the function, which, like a variable, can be called anything (provided it doesn't start with a number). It is best to keep your function names short and to the point. The purpose of this function is to trigger the sensor and return a distance measurement, so I've called this function get_distance().

Parentheses follow a function name, and the contents of such parentheses are referred to as the function's *parameters* or *arguments*. These parameters allow us to pass information into a function for later use. In our case, we pass the trigger and echo pin information we set up earlier, so that the function is able to activate and use the HC-SR04 distance sensor.

As with while and for loops, you need to indent the code inside the function so Python knows what code belongs to the function. The indented code begins at ❹ and extends to ❺, and it is *exactly* the same as the code you used to get a distance reading in Listing 5-1.

At ❺ the code *returns* the final output of the function: the distance reading. Returning information just means that the output of that function is handed back to the program whenever the function is called. This output could then be printed to the terminal, set to a variable, or manipulated in any way you, as the coder, desire!

We then start an infinite while loop. First we call the get_distance() function and store its result in the variable dist ❻.

Next, we introduce the crucial obstacle-avoiding logic with a conditional if statement ❼. This line translates to: "if the distance between the sensor and an obstacle is less than 15 cm, do the following." If the condition is true, the two lines inside the statement run and turn the robot slowly right for a quarter of a second.

Finally, the code at ❽ deals with any other scenario. If an obstacle is further than 15 cm away, the robot proceeds forward slowly for a tenth of a second. Obstacle-avoiding programs usually work better when the robot is moving at a slower speed, so we set the robot at 30 percent of its full speed with the (0.3) argument here. If you find this is too slow or too fast for your particular build, feel free to increase or decrease the values inside the parentheses of the motor commands.

Running Your Program: Make Your Robot Avoid Obstacles

Now that the final piece of code is finished, I recommend clearing a suitably large area and then strategically placing obstacles at the correct height for your ultrasonic distance sensor. Take a look at Figure 5-11 for the course I quickly created for my robot.

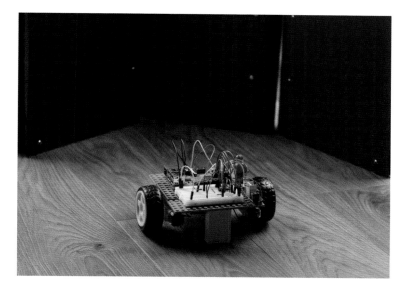

FIGURE 5-11

My robot facing down some looming upright folders

Run your program with the command:

```
pi@raspberrypi:~/robot $ python3 obstacle_avoider.py
```

Your robot should spring to life and proceed forward until it reaches its first obstacle, at which point it should turn until the obstacle is clear and then continue on its path.

Another fun experiment would be to stand in front of your robot and watch it scuttle away from you, no matter how many times you move your feet to be in front of it!

As ever, kill the robot with the command CTRL-C.

Challenge Yourself: Improve the Obstacle Avoidance Program

Our method of obstacle avoidance here leaves a lot of room for improvement!

As mentioned, the HC-SR04 is able to detect distance only in a single straight line, so your robot could miss obstacles that are directly in front of it but too low or too high for it to sense.

119 • CHAPTER 5

Having just one distance sensor is also a limitation. Your robot can detect only obstacles that are straight ahead, so it could easily turn right or left into another obstacle! The more distance sensors you use, the more information the robot has and therefore the more intelligent the running code can be.

Try to tweak the program in Listing 5-2 to make your robot avoid obstacles as efficiently as possible with one sensor. You could edit the minimum distance before evasive maneuvers are taken (the default is 15 cm). Or, you could edit the amount of time the motors are on for each evasion or the direction you turn. Try out different values in these variables and figure out what works best.

When you feel you've improved the program as much as you can, you could grab a second ultrasonic HC-SR04 module, wire it up to more of your Pi's GPIO pins just as before, and customize your code to use this new source of data in addition to the first HC-SR04. If you're using two sensors, a good place to mount them is on the front corners of your robot, rather than just facing forward.

If you're still feeling adventurous, you could try a third distance sensor to build an even better picture of the environment your robot is in!

SUMMARY

In this chapter we've covered everything from the theory behind ultrasonic distance measurement to programming using functions. You've put all of this together to turn your robot into a fully autonomous obstacle-avoiding machine.

In the next chapter, I'll show you how to make your robot unique by adding programmable RGB LEDs and sound effects!

6
CUSTOMIZING WITH LIGHTS AND SOUND

MAKING YOUR ROBOT STAND OUT FROM THE CROWD CAN BE A WHOLE LOT OF FUN. IN THIS CHAPTER, I'LL SHOW YOU HOW TO ADD LIGHTS AND SPEAKERS TO YOUR ROBOT TO MAKE IT FLASHIER, LOUDER, AND MORE EXCITING. AS USUAL, WE'LL COVER THE THEORY, THE PARTS YOU'LL NEED, AND HOW TO USE THEM.

ADDING NEOPIXELS TO YOUR RASPBERRY PI ROBOT

One of the best ways to grab attention is to have your robot put on a light show. With the right code and wiring and the help of some bright and colorful LEDs, you can make dazzling spectacles as your robot scuttles around the floor!

In this project you'll outfit your robot with a string of super-bright, multicolor LEDs. I'll guide you through getting the components, wiring them up, and programming different patterns. We'll combine these new additions with the Wiimote program from Chapter 4 so that you can trigger different LED combinations by pressing the Wiimote's buttons.

Introducing NeoPixels and the RGB Color System

At the start of this book, I introduced LEDs and showed you how to wire up a single-color LED to your Raspberry Pi and flash it on and off using a simple Python script.

That was a great project to get you started, but a lonely LED is hardly going to create the desired wow factor for your robot. Instead, for this project, we'll use *NeoPixels* like the ones shown in Figure 6-1.

FIGURE 6-1

NeoPixels on my robot

NeoPixels are a range of affordable, ultra-bright RGB LEDs from the open source hardware company Adafruit. *RGB*, which stands for *red green blue*, is a system of color mixing that computers use to represent a massive spectrum of colors. Red, green, and blue light can be combined in various proportions to produce any color in the visible light spectrum, from orange to indigo to green! By setting the levels of R, G, and B each in a range of 0 to 100 percent intensity, you can create new colors. For example, pure red is represented by 100% R, 0% G, and 0% B, and purple is 50% R, 0% G, and 50% B.

Instead of percentages, computers normally represent the levels of each color as a range of decimal numbers from 0 to 255 (256 levels). So, for red the combination is 255 R, 0 G, and 0 B. See Figure 6-2 for the full RGB range represented as a color wheel.

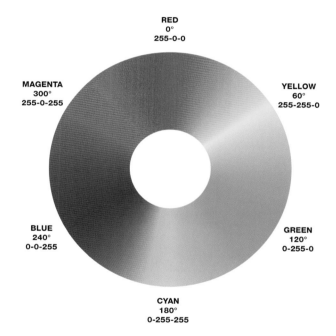

RGB COLOR WHEEL

RED
0°
255-0-0

MAGENTA
300°
255-0-255

YELLOW
60°
255-255-0

BLUE
240°
0-0-255

GREEN
120°
0-255-0

CYAN
180°
0-255-255

FIGURE 6-2

The full RGB range represented as a color wheel

This means that, unlike the single-color LED, each RGB NeoPixel can display a huge range of colors. You can calculate the exact range by multiplying the number of possibilities for each level: 256 × 256 × 256 = 16,777,216. That's almost 17 million different colors!

But how can a single LED represent so many colors? Well, if you look closely at the NeoPixel in Figure 6-3, you'll see that there are three distinct areas. This is because each NeoPixel actually comprises *three* LEDs: one each of red, green, and blue. You combine these colors in varying quantities, as discussed earlier, to produce an RGB color.

FIGURE 6-3

A macro shot of a
NeoPixel

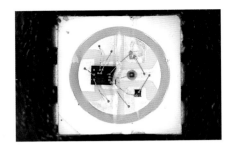

The Parts List

NeoPixels can be used individually or chained together, and Adafruit has a huge range of NeoPixel products available in many different forms and sizes—from individual pixels to huge matrices made out of hundreds of NeoPixels.

In this project I recommend picking up a NeoPixel *Stick*—this is a roughly 2-inch-long LED arrangement of eight NeoPixels, as shown in Figure 6-4. Its combination of small size and bright output makes it ideal for your robot.

FIGURE 6-4

NeoPixel Stick with
headers soldered on

If you're in the United States, you can buy one of these from Adafruit's website for less than $6. If you're elsewhere in the world, just search the net for "NeoPixel Stick," and you should have no trouble finding one for a similar price from another retailer.

It is worth noting that the NeoPixel Stick does require a small degree of assembly: you will have to solder a set of male *headers* onto the power and Data-In pads, as shown in Figure 6-5. There are two sets of pads on the back that look almost identical, except one side is for *input* into the Stick, and the other for *output* from the Stick. This is so you can chain the output of one Stick into the input of another Stick to join several together. We will only use one NeoPixel

Stick in this project, but you may want to experiment with more NeoPixels at a later point.

You'll have to purchase some male headers separately (which cost less than $1) and solder them to the set of the terminal pads that includes the *DIN* (Data-In) pin.

If you have never soldered before, check out "How to Solder" on page 204 for guidance.

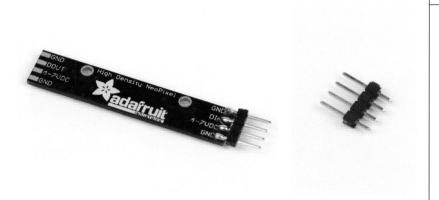

FIGURE 6-5

The back of my NeoPixel Stick with headers soldered to the input side (L); an individual four-pin header (R)

Other than the NeoPixel Stick and headers, you'll only need a few jumper wires to connect the NeoPixels, as well as some sticky tack to affix them to your Raspberry Pi robot.

Wiring Up Your NeoPixel Stick

Once you have a freshly soldered NeoPixel Stick, you can wire it up to your Pi. In total, only three connections are required to get it working. Remember that I won't show the previous connections in the diagrams, but you don't need to disconnect any of your previous projects to follow along with this one.

Like the HC-SR04 in the previous chapter, the Stick can be plugged directly into a breadboard, but as with that project, I don't recommend doing so here. Instead, it's better to connect the Stick using jumper wires so you can mount it elsewhere on your robot.

1. Use a jumper wire to connect the 4-7VDC pin of your NeoPixel Stick to the +5 V rail of your breadboard. Note that because these LEDs are extra bright, they draw a significant amount of current. Consequently, when we run the software later, you'll need to connect and switch on the robot's batteries.

2. Next, use another jumper wire to attach one of the GND pins of your Stick to the common ground rail of your breadboard. This grounds your NeoPixels to both the power supply (your batteries) and the Raspberry Pi. Check out Figure 6-6 for a diagram of what your setup should look like so far.

FIGURE 6-6

Adafruit NeoPixel Stick connected to +5 V and ground

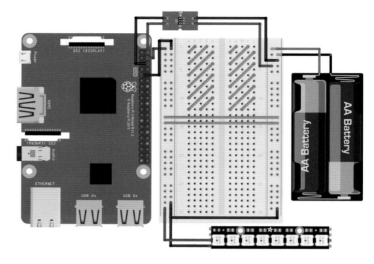

3. Use a final jumper wire to connect the DIN (Data-In) pin of your NeoPixel Stick to physical pin 19 (BCM 10) on your Raspberry Pi (see "Raspberry Pi GPIO Diagram" on page 200 for a guide to pin numbering). The complete circuit should look like Figure 6-7.

FIGURE 6-7

Complete breadboard diagram with the NeoPixel Stick wired up to power and your Pi

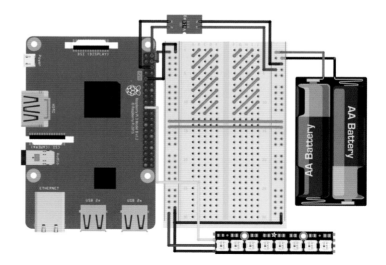

Use sticky tack to mount your NeoPixels somewhere on your robot. I've mounted mine to the right of the breadboard.

Installing the Software

Before you program your NeoPixel Stick, you must first install and configure the necessary software. The Python library we'll use is called `rpi_ws281x`, and you can download it from the internet for Python 3 using `pip`, a command line tool that allows you to quickly and easily install and manage Python software.

Before you proceed, you'll need to ensure that you have `pip` installed for Python 3. To do so, boot up your Raspberry Pi and log in via SSH. Then, enter the following command into the terminal:

```
pi@raspberrypi:~ $ sudo apt-get update
```

This command doesn't actually install new software; instead, it updates the list of available software your Raspberry Pi can download. After this process has completed, you can install `pip` for Python 3 with this command:

```
pi@raspberrypi:~ $ sudo apt-get install python3-pip
```

Most likely, you'll be informed that `pip` is already installed for Python 3, in which case you're ready to use it. If not, go through the installation process.

After this, you can install the `rpi_ws281x` library with one simple command:

```
pi@raspberrypi:~ $ sudo pip3 install rpi_ws281x
```

We will use the *SPI bus* to control the NeoPixels. This is just an electronics interface (the *serial peripheral interface*, to be exact) on select GPIO pins of every Raspberry Pi. By default, SPI is disabled, but you should have enabled it in the GUI when you set up your Pi at the start of this book. You can check that you have it enabled by opening the Raspberry Pi software configuration tool with this command:

```
pi@raspberrypi:~ $ sudo raspi-config
```

Once the tool opens, scroll down to **Interfacing Options**, select it, and then press ENTER. You'll be presented with the menu in Figure 6-8.

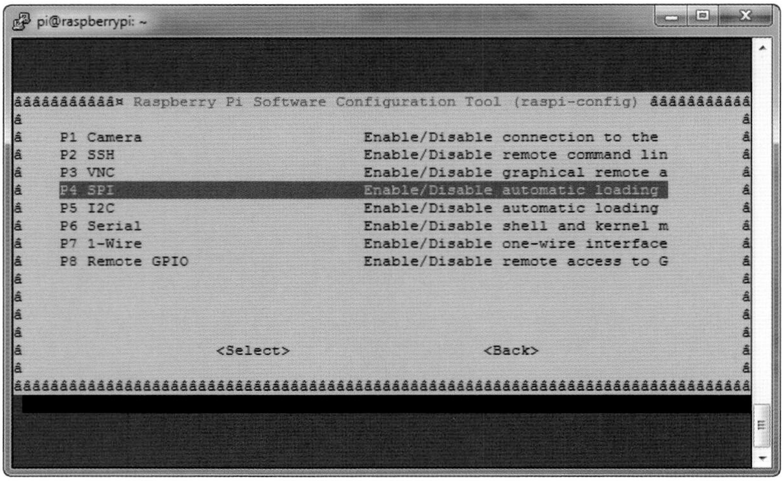

FIGURE 6-8

The Interfacing Options menu of the raspi-config tool

Scroll down and select **SPI**. You'll then be asked whether you would like the SPI interface to be enabled. Use the left/right arrow keys to highlight **Yes**. After you've done so, you'll return to the main raspi-config menu. You can exit the configuration tool by pressing the right arrow key twice (highlighting **Finish**) and then pressing ENTER. Now reboot your Raspberry Pi using this command:

```
pi@raspberrypi:~ $ sudo reboot
```

Now SPI is enabled!

MAKING SPI WORK CORRECTLY ON PI 3

If you're using the Raspberry Pi 3 Model B/B+ there is one more step you'll have to take before moving on. You don't have to worry about this step if you're using an older Pi.

To get SPI to work correctly on a Pi 3, you'll have to change the GPU core frequency to 250 MHz. All this means is that you are changing the graphics unit on your Raspberry Pi 3 to run at a slightly different rate. If you don't do this, your NeoPixels may behave erratically and not display the correct patterns.

To make this change, enter the following command into the terminal:

```
pi@raspberrypi:~ $ sudo nano /boot/config.txt
```

This will open up a configuration file containing various text and options. Scroll down to the bottom of this file and, on a new line, add this text:

```
core_freq=250
```

So, for example, the end of my configuration file looks like this:

```
--snip--
# Additional overlays and parameters are documented /boot/
overlays/README

# Enable audio (loads snd_bcm2835)
dtparam=audio=on
start_x=1
gpu_mem=128

core_freq=250
```

Once you have added this line, save the file in Nano by pressing CTRL-X, and then press Y and ENTER. Then reboot your Raspberry Pi.

NOTE

If you think that this process may have changed, or you're concerned that you have done this step incorrectly, check the book's website at https://nostarch.com/raspirobots/.

Configuring the Library's Example Code

Before we go any further, let's test the library you just installed to make sure everything is working perfectly. If you have already downloaded the software bundle that comes with this book onto your Raspberry Pi, then you already have the test file, *strandtest.py*. This program has been written by Adafruit to test out NeoPixels. If you don't have it, download the example code from the internet by entering the following command:

```
pi@raspberrypi:~/robot $ wget https://raw.githubusercontent.com/
the-raspberry-pi-guy/raspirobots/master/strandtest.py
```

After this has finished, you will have the exact same test code that is provided in the software bundle.

Before running the example code, we need to change a few settings. To look at the code and the current settings, open the example code file using Nano as follows:

```
pi@raspberrypi:~/robot $ nano strandtest.py
```

The purpose of this program is run through several example light patterns. The code is quite long and full of functions that define the different sequences, but you don't need to edit any of this code.

You will, however, need to edit some of the constants inside the program. Near the start of the code you'll find the chunk of code to be edited, which is reproduced in Listing 6-1.

LISTING 6-1

LED strip configuration
of *strandtest.py*

```
# LED strip configuration:
❶ LED_COUNT      = 16        # Number of LED pixels.
❷ LED_PIN        = 18        # GPIO pin connected to the pixels ↵
                             (18 uses PWM!).
❸ #LED_PIN       = 10        # GPIO pin connected to the pixels ↵
                             (10 uses SPI /dev/spidev0.0).
  LED_FREQ_HZ    = 800000    # LED signal frequency in hertz (usually ↵
                             800khz)
  LED_DMA        = 10        # DMA channel to use for generating ↵
                             signal (try 10)
  LED_BRIGHTNESS = 255       # Set to 0 for darkest and 255 for ↵
                             brightest
  LED_INVERT     = False     # True to invert the signal (when using ↵
                             NPN transistor level shift)
  LED_CHANNEL    = 0         # set to '1' for GPIOs 13, 19, 41, 45 ↵
                             or 53
  LED_STRIP      = ws.WS2811_STRIP_GRB   # Strip type and color ↵
                                         ordering
```

The words after each hash character (#) are *comments*. Programmers will often put comments in their code as a form of annotation. Comments help human readers and other programmers understand what the various parts of your program do.

In Python, a comment starts with a hash character (#). When Python interprets this code, it simply ignores everything after the hash. It is good coding practice to comment your programs, especially if you are working in a team or open-sourcing your work. Commenting is also handy as a reminder to yourself if you revisit a program in the future and have forgotten how it works!

The first thing you need to change appears at ❶: LED_COUNT. This is a constant for the number of NeoPixels you have attached to your Pi. By default it is set to 16, so you need to change it to 8 instead.

Secondly, you'll change the pin number being used. The constant LED_PIN at ❷ is set to BCM 18 by default, but your NeoPixel Stick is connected to BCM 10. The authors of this example code have noticed that using BCM 10 is a popular choice, so they've provided an alternative constant definition at ❸, but commented it out.

To swap these lines around, add a hash to the start of the line at ❷. This will *comment out* that line so that Python will ignore it. Then, remove the hash at ❸ to *uncomment* the line, which will make Python run the line of code assigning LED_PIN to 10.

Your final block of constants should now look like the code in Listing 6-2.

```
# LED strip configuration:
LED_COUNT      = 8        # Number of LED pixels.
#LED_PIN       = 18       # GPIO pin connected to the pixels (18 ↵
                            uses PWM!).
LED_PIN        = 10       # GPIO pin connected to the pixels (10 ↵
                            uses SPI /dev/spidev0.0).
LED_FREQ_HZ    = 800000   # LED signal frequency in hertz (usually ↵
                            800khz)
LED_DMA        = 10       # DMA channel to use for generating ↵
                            signal (try 10)
LED_BRIGHTNESS = 255      # Set to 0 for darkest and 255 for ↵
                            brightest
LED_INVERT     = False    # True to invert the signal (when using ↵
                            NPN transistor level shift)
LED_CHANNEL    = 0        # set to '1' for GPIOs 13, 19, 41, 45 or 53
LED_STRIP      = ws.WS2811_STRIP_GRB   # Strip type and color ↵
                                        ordering
```

LISTING 6-2

The new LED strip configuration of *strandtest.py*

Once you have your code set up, you can run it.

Running the Example Code

Save the changes you have made to the example program and run it with the following command:

```
pi@raspberrypi:~/robot $ python3 strandtest.py -c
```

And you might want to grab a pair of sunglasses! Your NeoPixels should now be going through a sequence of patterns, the names of which are displayed in the terminal as they start (see Figure 6-9).

FIGURE 6-9

NeoPixels going through the *strandtest.py* example program

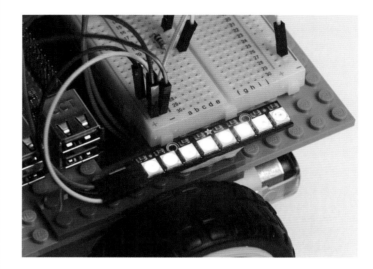

When your retinas have had enough, press CTRL-C to kill the example code. The -c you should have added to the end of your original run command should turn off your LEDs. If you didn't add -c to the command, killing the program will just freeze your LEDs and they will remain powered on.

If you're worried about the intensity of your LEDs and being blinded, don't fret! As we add NeoPixel control to the Wiimote program, I'll show you how to turn down their brightness.

Controlling NeoPixels Using the Wiimote Program

Now that you have tested out your NeoPixels and seen what they're capable of, it's time to add LED-controlling capabilities to the accelerometer-based Wiimote program you made earlier.

We're going to add NeoPixels to the Wiimote program, but it would be ideal to keep a copy of the original Wiimote code without these additions just in case something goes wrong or we want to go back to it in future. To do this, we'll create and edit a copy of the program instead. First, make sure you are in the directory where your code is stored; for me, that's my *robot* directory. Then, in the terminal, copy your original Wiimote program using the command `cp`:

```
pi@raspberrypi:~/robot $ cp remote_control_accel.py neo_remote_
control.py
```

This command simply copies the contents of its first argument (*remote_control_accel.py*) to the new file specified in the second argument. As you can see, I have decided to name my NeoPixel version of the Wiimote program *neo_remote_control.py*. After this, open the newly copied file with Nano like so:

```
pi@raspberrypi:~/robot $ nano neo_remote_control.py
```

Now enter the modifications to the code in Listing 6-3, or you can download the complete program at *https://nostarch.com/raspirobots/*. I have omitted and compressed the parts of the program that have not changed.

```
   import gpiozero
   import cwiid
   import time
❶ from rpi_ws281x import *

   robot = gpiozero.Robot(left=(17,18), right=(27,22))
   --snip--
   wii.rpt_mode = cwiid.RPT_BTN | cwiid.RPT_ACC

   LED_COUNT      = 8
   LED_PIN        = 10
   LED_FREQ_HZ    = 800000
   LED_DMA        = 10
❷ LED_BRIGHTNESS = 150
   LED_INVERT     = False
   LED_CHANNEL    = 0
   LED_STRIP      = ws.WS2811_STRIP_GRB

❸ strip = Adafruit_NeoPixel(LED_COUNT, LED_PIN, LED_FREQ_HZ,
   LED_DMA, LED_INVERT, LED_BRIGHTNESS, LED_CHANNEL, LED_STRIP)
   strip.begin()

❹ def colorWipe(strip, color, wait_ms=50):
       """Wipe color across display a pixel at a time."""
❺     for i in range(strip.numPixels()):
           strip.setPixelColor(i, color)
           strip.show()
           time.sleep(wait_ms/1000.0)
```

LISTING 6-3

The updated Wiimote code with NeoPixel functionality

```
   while True:
❻     buttons = wii.state["buttons"]
      if (buttons & cwiid.BTN_PLUS):
          colorWipe(strip, Color(255, 0, 0))  # Red wipe
      if (buttons & cwiid.BTN_HOME):
          colorWipe(strip, Color(0, 255, 0))  # Blue wipe
      if (buttons & cwiid.BTN_MINUS):
          colorWipe(strip, Color(0, 0, 255))  # Green wipe
      if (buttons & cwiid.BTN_B):
          colorWipe(strip, Color(0, 0, 0))    # Blank

      x = (wii.state["acc"][cwiid.X] - 95) - 25
      --snip--
      if (turn_value < 0.3) and (turn_value > -0.3):
          robot.value = (forward_value, forward_value)
      else:
          robot.value = (-turn_value, turn_value)
```

This program relies on two additional sets of libraries that the original Wiimote code did not, so we need to import both the `time` and `rpi_ws281x` libraries ❶.

Then, just as with the original program, we set up the robot and Wiimote for use. After this, we define the same chunk of constants we saw in the example NeoPixel program. These define the various parameters for the NeoPixel Stick. Most notably, you'll find `LED_BRIGHTNESS` ❷, a constant that can be set between 0 and 255. I have set mine to be dimmer and easier on the eyes at 150.

At ❸, we create the NeoPixel Stick object and set up the constants defined previously. The library is initialized on the following line.

We then define a function called `colorWipe()` ❹ to use later. This function has been taken directly out of the *strandtest.py* example. The comment inside it describes what the function does: it wipes a color across the NeoPixel Stick one pixel at a time. To do this, it takes an RGB `color` parameter and then uses a `for` loop ❺ to set each pixel to that color one by one, with a short delay in between.

After this, we start the main body of code in the infinite `while` loop. At the start of each loop, the status of the Wiimote buttons is read ❻. Then, depending on whether the user presses the plus, minus, or home button, a different color will wipe across the NeoPixel Stick and remain there until another button is pressed. If the user presses the B button, the NeoPixels will be reset.

The rest of the program is exactly the same as the original: it deals with the accelerometer output from the controller and makes the robot move accordingly.

Running Your Program:
NeoPixels and Wiimote Control

Save your work and run your code with the following command:

```
pi@raspberrypi:~/robot $ python3 neo_remote_control.py
```

Your robot should now respond to the accelerometer data from your Wiimote. Try pressing the plus, minus, home, and B buttons to trigger the different lights, as shown in Figure 6-10.

FIGURE 6-10

My robot with its NeoPixels set to blue

Before you kill the program with CTRL-C, make sure you press the B button on your Wiimote to turn the NeoPixels off!

Challenge Yourself:
Experiment with Color and Pattern

Once you have played around with your robot and NeoPixels, go back into the program and the example code shown previously to see if you can set your own custom colors by changing the RGB color combinations. Or, see if you can create more adventurous light patterns to display.

If you have more than one NeoPixel Stick, you can chain them together by feeding the output of one into the input of the other to create an even more dazzling two-wheeler!

ADDING A SPEAKER TO YOUR RASPBERRY PI ROBOT

WARNING

You'll only be able to follow along with these projects if you have a full-size Raspberry Pi like the Pi 3, Pi 2, Pi 1 Model B/B+ or even A+. Models such as the Pi Zero and Pi Zero W do not feature 3.5 mm audio jacks and therefore can't easily connect to a speaker.

While your robot has already come a long way, one feature that has been notably absent is the ability to make noise and communicate. In the following two projects, we'll change that! I'll guide you through adding a small 3.5 mm speaker to your robot and using it to add sound to two previous projects: a car horn for the Wiimote program, and a parking-sensor-style beep for your obstacle avoidance program.

Understanding How 3.5 mm Speakers Work

A *loudspeaker* (or just plain *speaker*) converts an electrical audio signal into a sound that can be heard by humans. You'll have come across many speakers in a wide range of environments—from huge speakers at concerts to the minuiscule ones inside your mobile phone.

In order to translate an electrical signal into an audible sound, speakers use an electromagnet to vibrate a cone. This cone amplifies those vibrations and pumps sound waves into the surrounding air and to your ears.

For the following two projects you'll need a small 3.5 mm speaker, like the one shown in Figure 6-11. The 3.5 mm sizing refers to the diameter of the audio jack. This size is an industry standard, and the same as most phone headphone jacks.

FIGURE 6-11

My small 3.5 mm speaker

You can pick up a 3.5 mm speaker online by searching eBay, Amazon, or any regular electronics retailer. It should set you back no more than $10. The make and brand isn't that relevant; as long as it is small enough to fit on your robot, is relatively loud, and has a 3.5 mm jack, you should be good!

Connecting Your Speaker

Most small speakers are rechargeable, so before you begin connecting your speaker to your Raspberry Pi, make sure that it is fully charged and operational.

Your Pi's 3.5 mm audio jack is between the HDMI and Ethernet ports. Take your speaker and plug its 3.5 mm cable into the jack on your Pi, as shown in Figure 6-12.

NOTE

If your speaker isn't rechargeable, the method you use to power it will depend on the exact model. If it requires USB power, you could plug it into one of your Pi's USB ports. Having a rechargeable speaker avoids this issue and is therefore the most ideal option in this situation.

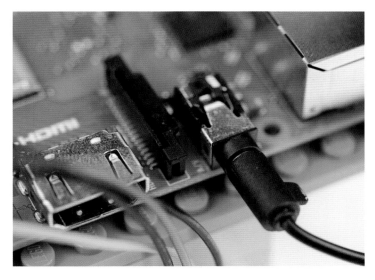

FIGURE 6-12
Speaker connected to my Raspberry Pi through the 3.5 mm audio jack

Now mount your speaker somewhere on your robot chassis. Where you mount it depends on the size of your speaker, as well as the free space that you have available. Since I didn't have enough room to simply attach the speaker to the main chassis, I decided to create a small stalk out of a few LEGO pieces, and then affixed my speaker with sticky tack, as shown in Figure 6-13.

FIGURE 6-13

My 3.5 mm speaker
connected to my Pi and
mounted on top of a
small LEGO stalk

ADDING A CAR HORN TO THE WIIMOTE PROGRAM

Now let's extend the previous program so that your robot sounds a car horn at your command. We'll edit the NeoPixel Wiimote program to activate a horn sound effect when the A button on your Wiimote is pressed.

Installing the Software

Normally you'd play an audio file by clicking it on a GUI and opening it in a music playing application. Unlike a GUI, though, the terminal leaves you with no such ability, so you have to use special commands to play audio files. As with the NeoPixels, you first need to install the required software and configure the sound output.

First, ensure that the `alsa-utils` software package is already installed on your Raspberry Pi. This is a collection of software that relates to audio and device drivers. You can check whether it's installed or install the package using this command:

```
pi@raspberrypi:~/robot $ sudo apt-get install alsa-utils
```

If your Pi tells you that it already has the latest version of `alsa-utils`, then great! If not, you'll need to go through the quick installation process, responding to the prompts.

After this, the only remaining step is to tell the Raspberry Pi to play audio through the 3.5 mm audio jack instead of the HDMI port. We do this in the terminal by using the Raspberry Pi configuration tool `raspi-config`, just like we did earlier. To open this tool, use the following command:

```
pi@raspberrypi:~/robot $ sudo raspi-config
```

You should see a blue screen with options in a gray box in the center, as in Figure 6-14.

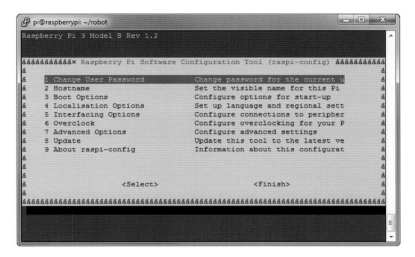

FIGURE 6-14

The Raspberry Pi software configuration tool

Now use the arrow keys to scroll down and select **Advanced Options** and then press ENTER. This will open up a new menu; scroll down to **Audio**, select it, and press ENTER again.

Once here, you will be provided with the three options. Select the **Force 3.5mm jack** option, as shown in Figure 6-15.

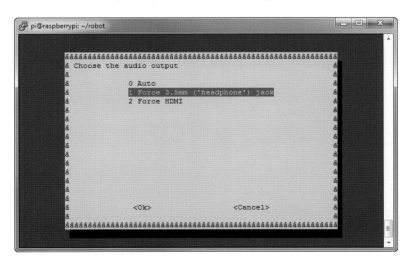

FIGURE 6-15

Choosing the audio output using `raspi-config`

Next, you'll be returned to the original menu shown in Figure 6-14. From there exit the configuration tool by pressing the right arrow key twice (to highlight **Finish**) and then pressing ENTER.

Playing Sounds from the Terminal

To play sounds from the terminal, you first need some sounds to play! The audio files for this project and the next one can be found online at *https://nostarch.com/raspirobots/*. If you have downloaded all of the software in bulk, then you'll already have the files. Alternatively, you can grab the two audio files off the internet with a few easy commands. Either way, first create a new directory called *sounds* inside the folder where you're storing all of your robot programs. For me, this command looks like:

```
pi@raspberrypi:~/robot $ mkdir sounds
```

If you downloaded the files in bulk, transfer the files *beep.wav* and *horn.wav* into this new folder. If you want to download the files directly, then change into that directory as follows:

```
pi@raspberrypi:~/robot $ cd sounds
```

Finally, to download each of the audio files, use this command:

```
pi@raspberrypi:~/robot/sounds $ wget https://github.com/the-raspberry-pi-guy/raspirobots/tree/master/Chapter%206%20-%20Adding%20RGB%20LEDs%20and%20Sound/sounds/beep.wav
```

Followed by this one:

```
pi@raspberrypi:~/robot/sounds $ wget wget https://github.com/the-raspberry-pi-guy/raspirobots/tree/master/Chapter%206%20-%20Adding%20RGB%20LEDs%20and%20Sound/sounds/horn.wav
```

Now if you enter ls in the terminal, you'll find two new audio files—*horn.wav* and *beep.wav*:

```
pi@raspberrypi:~/robot/sounds $ ls
horn.wav  beep.wav
```

The former is the file we'll use in this project. Before you test *horn.wav*, increase the software volume level of your speaker to its maximum with this command:

```
pi@raspberrypi:~/robot/sounds $ amixer set PCM 100%
```

Also ensure that any physical volume control on your 3.5 mm speaker is at its maximum. Then, to play *horn.wav* through your 3.5 mm speaker, you'll use `aplay`, a terminal-based sound player, like so:

```
pi@raspberrypi:~/robot/sounds $ aplay horn.wav
Playing WAVE 'horn.wav' : Signed 24 bit Little Endian in 3bytes,
Rate 44100 Hz, Stereo
```

You should hear your robot emit a single car horn noise!

Playing Sound Using the Wiimote Program

Now that you understand how to play sound files through the terminal, you can add this functionality to the Wiimote program from earlier in the chapter. This means that your robot will be able not only to trigger a light show, but also to sound a car horn whenever you wish!

To accomplish this, we'll call the `aplay` command from inside Python. Navigate back into the *robots* directory and then reopen the NeoPixel/Wiimote code with this command:

```
pi@raspberrypi:~/robot $ nano neo_remote_control.py
```

Then, enter the additions in Listing 6-4 into your own code. As before, all of the unchanged code has been omitted. Alternatively, you can grab the modified file from the book's website.

```
import gpiozero
import cwiid
import time
from rpi_ws281x import *
❶ import os

robot = gpiozero.Robot(left=(17,18), right=(27,22))
--snip--

while True:
    buttons = wii.state["buttons"]
    if (buttons & cwiid.BTN_PLUS):
        colorWipe(strip, Color(255, 0, 0))  # Red wipe
    --snip--
    if (buttons & cwiid.BTN_B):
        colorWipe(strip, Color(0, 0, 0))    # Blank

❷    if (buttons & cwiid.BTN_A):
        os.system("aplay sounds/horn.wav")

    x = (wii.state["acc"][cwiid.X] - 95) - 25
    --snip--
```

LISTING 6-4
The modified NeoPixel/Wiimote code with car horn sound effect

```
if (turn_value < 0.3) and (turn_value > -0.3):
    robot.value = (forward_value, forward_value)
else:
    robot.value = (-turn_value, turn_value)
```

The additions required are simple and span only three lines. The first thing to note is at ❶, where the os library is imported. The os library enables us to use the functionality of the Pi's operating system inside of a Python program.

This comes in handy at ❷. Here, the program detects whether the user has pressed the A button on the Wiimote. If so, the same aplay terminal command you used earlier is called using os.system. Notice that there is also a short filepath to the *horn.wav* sound, as this file is stored in a different directory than the program.

Running Your Program: NeoPixels, Sound Effects, and the Wiimote Control

Save your work and run it with the same command as the previous project:

```
pi@raspberrypi:~/robot $ python3 neo_remote_control.py
```

Your robot will now respond exactly as before, with accelerometer control. You'll also be able to trigger the same lights as before. Now try pressing the A button: you should hear your robot honk its horn!

ADDING BEEPING TO THE OBSTACLE AVOIDANCE PROGRAM

In this project, we'll revisit the obstacle avoidance program you already coded in Chapter 5, and add a beeping sound to alert you when your robot has detected an obstacle within a 15 cm range.

Integrating the Beep Sound into the Obstacle Avoidance Program

You've already set up your speaker and configured the necessary software, so we can jump straight into integrating the beep sound into the obstacle avoidance program.

We'll do this as we did with the horn: by calling aplay inside of the Python program. I recommend using cp to create a new copy of the obstacle avoidance program. I've called mine *beep_obstacle_avoider.py*. Enter the modifications I made as shown in Listing 6-5.

LISTING 6-5

The *beep_obstacle_avoider.py* program

```
import gpiozero
import time
❶ import os

TRIG = 23
ECHO = 24

trigger = gpiozero.OutputDevice(TRIG)
--snip--

while True:
    dist = get_distance(trigger,echo)
    if dist <= 15:
❷        os.system("aplay sounds/beep.wav")
        robot.right(0.3)
        time.sleep(0.25)
    else:
        robot.forward(0.3)
        time.sleep(0.1)
```

Just as before, we import the os module ❶. Then, if the sensor detects an object less than or equal to 15 cm away, the program plays the beep sound ❷ and the robot changes course.

Running Your Program: Beeping Obstacle Avoidance

Save your work and run it with the following command:

```
pi@raspberrypi:~/robot $ python3 beep_obstacle_avoider.py
```

Your robot will now avoid obstacles and beep when it does so!

Challenge Yourself: Add Sound Effects to Your Other Projects

Now that you know the relatively simple process of adding audio effects to a program, why not revisit the other programs you have written over the course of this book and add sound to them? You could use your phone to record your own noises, or you can use online sound libraries that provide free audio files. For example, take a look at Freesound: *https://freesound.org/*.

SUMMARY

In this chapter you have decked out your robot with some super-bright NeoPixels and given it the gift of sound, too! Over the course of three different projects, we've covered everything from the theory of RGB LEDs to how to play audio in the terminal.

In the next chapter, we'll make your robot a little bit more intelligent! I'll guide you through the process of giving your two-wheeler the ability to autonomously follow lines.

7
LINE FOLLOWING

IN THIS CHAPTER YOU'LL GIVE YOUR ROBOT THE ABILITY TO DETECT AND FOLLOW A LINE AUTONOMOUSLY. IT WILL BE ABLE TO IDENTIFY ITS OWN COURSE AND THEN DECIDE EXACTLY HOW TO STICK TO IT.

This is a test of both digital recognition and clever programming. It is also a classic robotics task that is important for everyone to master, from beginners to professionals. By the end of this project you'll have a fully autonomous robot that will be able to stick to a line like glue!

CREATING A TRACK

As usual, prior to launching into any project, it is important and useful to step back and analyze the task at hand. The aim here is to make your robot follow a line. More specifically, you want to make your robot follow a black line on a white background, like the one shown in Figure 7-1.

FIGURE 7-1

My robot following a
black line

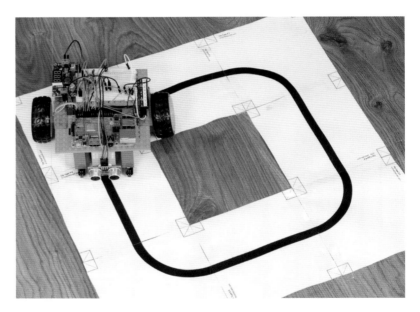

The combination of white and black provides maximum contrast for your robot, enabling us to use simple line-following sensors.

The first thing to do is create a line for your robot to follow. You can make the shape of this track as adventurous as you like. The simplest option is a basic loop, but feel free to get creative, as long as the background is white and the line is black. I also recommend making the width of the line around 1/4 inch.

There are many different ways of constructing a track. You could simply grab a large piece of paper (at least ledger size, roughly 11 × 17 inches) and draw a thick line with a black marker. You could use some black electrical tape on white poster board. You can even purchase premade line-following tracks online. If you have access to a printer, I recommend printing a track out on letter-size paper and assembling it with tape.

I've also included the template for a set of tiles you can make into a track of any shape in the resources for this book, which you can access on your Windows, Mac, or Linux PC:

1. If you haven't done so already, download the software bundle from *https://nostarch.com/raspirobots/* onto your personal computer.

2. Navigate to the folder where the software is stored, and then open the PDF named *track_generator.pdf*. This is a 34-page document that has a variety of 20 × 20 cm tiles with different lines on them (see Figure 7-2).

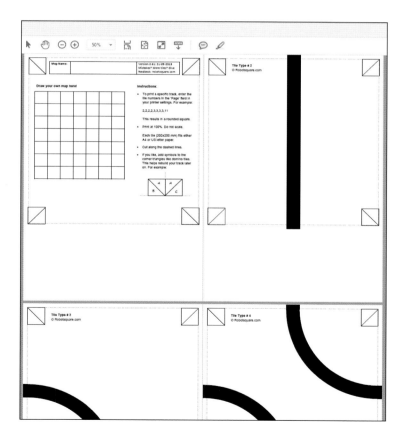

FIGURE 7-2

The *track_generator.pdf* document

3. Print out whichever lines you want to put together into a track, cut them out, and then stick them together to form your own custom track. The first few pages of the PDF document show simple paths, such as straight lines and corners, but the farther down you scroll, the crazier the paths get!

   For your first track, I recommend keeping it relatively simple. Remember that you can always print out more tiles and make a more difficult course for your robot in the future.

4. To create a simple loop, like the one in Figure 7-3, print out four copies of Tile Type #2 (the straight line) and four copies of Tile Type #3 (the basic corner). You should be able to print these pages specifically using the printer dialog box of your PDF reader software. Ensure that each tile takes up one letter-size piece of paper.

5. Use scissors to cut around the dotted line of each tile. Then arrange them in a loop, as shown in Figure 7-3, and use tape along the length of the *underside* to join the edges of the tiles together. Your track is now complete!

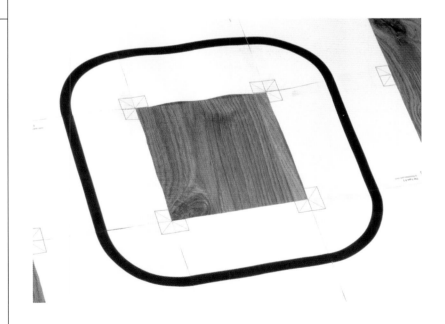

THE THEORY BEHIND LINE FOLLOWING

We're going to use *infrared* (IR) sensors to make your Raspberry Pi robot follow the black line. We used similar technology for obstacle avoidance in Chapter 5, when we used ultrasound for the purpose of detecting objects. But instead of sound, in this chapter we'll use invisible light. Fortunately, much of the same theory you learned previously can be applied here too.

Every IR sensor, like the one in Figure 7-4, has two small bulb-like devices—an infrared transmitter and a receiver, normally arranged and positioned closely together. The transmitter is an IR LED that, when triggered, fires off a pulse of infrared light. The receiver, an IR photodiode, then waits for this transmitted light to return. A *photodiode* is simply a device that uses light to vary an electric current.

FIGURE 7-4

An IR sensor

Light interacts differently with different types of surfaces. Most notably, light is *reflected* more by a white surface, and almost totally *absorbed* by a black surface, enabling the IR sensor to detect black lines on white backgrounds.

As depicted in Figure 7-5, if an IR sensor module is over a white surface, the receiver will detect the reflection of the infrared beam emitted by the transmitter. If the sensor is over a black surface, like the line of your track, the receiver will not detect the reflection. This difference in reflectance allows the sensor module to detect whether or not there is a line in front of it.

FIGURE 7-5

The different behavior
of IR light as it meets
a white versus a black
surface

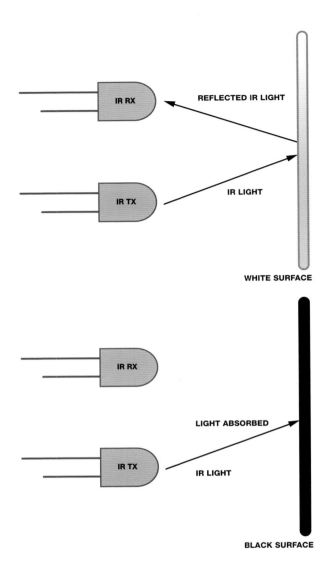

Mounting a single IR sensor onto the bottom of your robot would enable it to detect the presence of a black line, but if your robot moved so the sensor was no longer directly over the line, it could easily go off-track. With only one sensor, there is no easy way for your robot to detect whether it has gone too far to the left or right of the line. Instead, we'll use two IR sensors, both mounted on the bottom of your robot at the front, about an inch apart. Two sensors will provide a feedback mechanism with a sense of direction. There are four possibilities for the outputs of these sensors, each of which will guide the robot:

- If both sensors receive a reflected signal and detect white, the robot can assume that the line must be in between the sensors. Therefore, the robot should move forward in a straight line (see Figure 7-6).

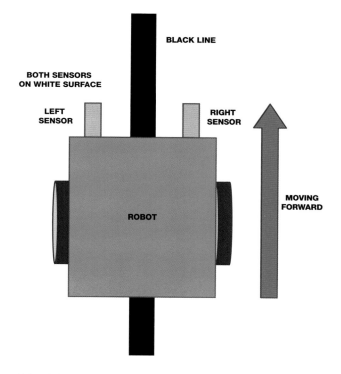

FIGURE 7-6

The robot moves forward when both sensors detect white.

- If the left sensor doesn't receive a reflected signal but the right sensor does, this must mean that the left sensor detects the line. This indicates that the robot must be veering right, so it should turn left to correct itself, as in Figure 7-7.
- If the right sensor detects the line but the left sensor doesn't, the robot should correct itself by turning right, as shown at the bottom of Figure 7-7.

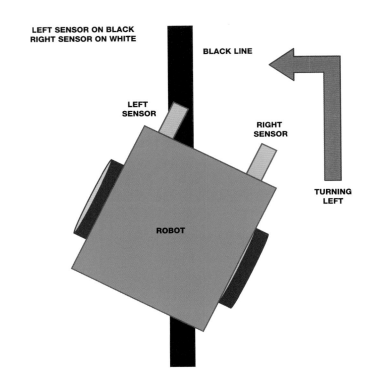

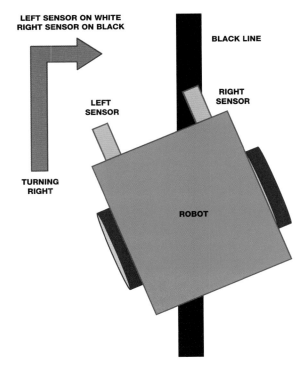

FIGURE 7-7

The robot turns left or right if one of the sensors detects the black line.

LEFT SENSOR ON BLACK
RIGHT SENSOR ON WHITE

BLACK LINE

LEFT
SENSOR

RIGHT
SENSOR

TURNING
LEFT

ROBOT

LEFT SENSOR ON WHITE
RIGHT SENSOR ON BLACK

BLACK LINE

LEFT
SENSOR

RIGHT
SENSOR

TURNING
RIGHT

ROBOT

- Finally, if neither sensor receives a reflected signal, then they must *both* be reading black, as shown in Figure 7-8 (this won't happen if you're using just a basic loop track). What to do next is up to you, but one option is simply to make the robot stop. If you experiment with other track layouts—for example, a figure 8 shape—you might run into other situations where both sensors detect black. In these circumstances you may want your robot to move forward, turn, or even reverse—play around to find out what works best!

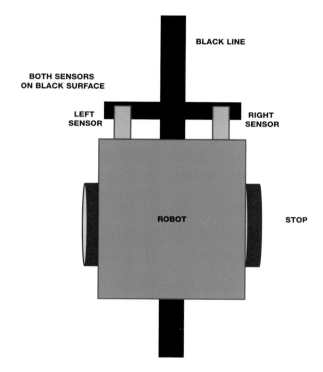

BLACK LINE

BOTH SENSORS ON BLACK SURFACE

LEFT SENSOR

RIGHT SENSOR

ROBOT

STOP

USING AN IR SENSOR TO DETECT A LINE

Before you start mounting two IR sensors to your Raspberry Pi and configuring the code behind a complete line-following robot, let's wire up just one IR sensor and test its line detection response.

The Parts List

For this part of the project you'll use only one sensor, but make sure to buy two, as you'll need both for the next project!

- 2 TCRT5000-based infrared line-following sensor modules
- Jumper wires

TCRT5000-based line-following sensor modules, like the one shown in Figure 7-9, are very common and available online for less than a couple of dollars each. I sourced mine from eBay by searching "TCRT5000 Line Follow Module." The *TCRT5000* part of the sensor name refers to the infrared optical sensor, the small black component on the underside of the board.

FIGURE 7-9

My TCRT5000 infrared line-following sensor module

Make sure to get a line-following *module* like the one in Figure 7-9, as this means the functionality of the optical sensor has been neatly packaged into an easy-to-use component. These boards have a simple set of pins and require just three connections (see Figure 7-10).

FIGURE 7-10

Pinout of a TCRT5000 IR sensor module

The infrared diode is housed inside the small black component and emits an infrared ray continuously when the module is connected to power. If the light is not reflected back into the sensor, it means a black line must be present and the output pin (OUT) of the module goes low (that is, it drops the voltage). This simple digital logic is ideal and, better still, the module can be powered natively from the Pi's 3.3 V. This saves us from having to use a voltage divider circuit as we did for the ultrasonic distance sensor in Chapter 5.

Wiring Up Your TCRT5000 Line-Following Sensor Module

Disconnect your Pi from power and wire up your sensor by following these instructions:

NOTE

Remember that you don't need to disconnect any of your previous circuits, but I won't show those connections here.

1. Use an F-F jumper wire to connect the VCC pin of your TCRT5000 module to physical pin 1 of your Raspberry Pi. This is the +3.3 V connection and provides power to the sensor.

2. Use another jumper wire to connect the GND pin of your module to the common ground rail on your breadboard. So far, your wiring should look like Figure 7-11.

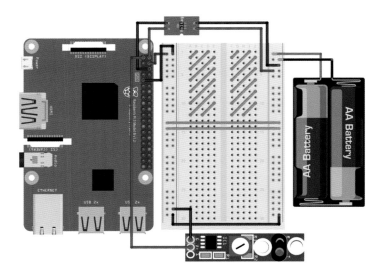

FIGURE 7-11

Line-following sensor connected to +3.3 V and GND

3. Use another jumper wire to connect the data output pin of your sensor (OUT) to physical pin 21 on your Raspberry Pi. This is BCM 9. Your complete circuit should look like Figure 7-12.

FIGURE 7-12

Completed
breadboard diagram
with your TCRT5000
line-following sensor
module wired up and
in place

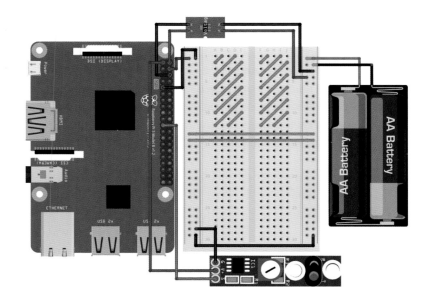

Programming Your Raspberry Pi to Detect a Line

Now that you have a line sensor connected to your Raspberry Pi,
let's write some code to test its line detection response. Connect
your Pi to power via a wall outlet, boot it up and move into your
code folder, and then enter the following to create and open a test
program called *line_test.py*:

```
pi@raspberrypi:~/robot $ nano line_test.py
```

Add the code in Listing 7-1, which will test your TCRT5000 module. The purpose of this program is to simply output to the terminal to
inform the user whether the module has detected a line.

LISTING 7-1

Detecting a line

```
import gpiozero
import time

❶ line_sensor = gpiozero.DigitalInputDevice(9)

while True:
    ❷ if line_sensor.is_active == False:
            print("Line detected")
    ❸ else:
            print("No line detected")

    time.sleep(0.2)
```

After importing the usual libraries, we set up the line sensor as a digital input on BCM 9 ❶.

We then start an infinite `while True` loop containing the logic of the program. At ❷ we use an `if` statement to detect whether the line sensor is active. If it *isn't* active, it must mean the infrared reflection has not returned and therefore the emitted rays have been absorbed by the black line. Thus, the line has been detected, and we indicate this to the user with the `print()` statement inside the `if` statement.

Any other scenario would mean that the line sensor is active and therefore is *not* detecting a line. The `else` statement ❸ catches this alternative scenario and outputs to the user that a line has not been detected. The program then waits for a fifth of a second and loops back around.

NOTE

The TCRT5000 has a detection distance between 1 mm and 8 mm. If the line is farther away from the sensor than this, it is likely to give false positives or negatives.

Running Your Program: Detect a Line!

After saving your program, grab a piece of your track to test out your sensor. To run the program, enter:

```
pi@raspberrypi:~/robot $ python3 line_test.py
```

While there is nothing within range of the sensor, it may produce erratic results that will be displayed in the terminal. Bring your piece of track up to the module and move the line over the sensor. You should notice that as you move the piece, the terminal output will change to `Line detected`:

```
pi@raspberrypi:~/robot $ python3 line_test.py
No line detected
No line detected
Line detected
Line detected
```

My particular sensor also has an LED on the top that changes state when the line comes into view.

If you're having issues with your module and it's not detecting the line very well, you can try a few things. First, try to limit interference from other light sources by turning the lights off. Alternatively, some TCRT5000 modules have an on-board potentiometer to adjust their sensitivity. This usually looks like a piece of blue/white plastic with a place to insert a screwdriver. Use an appropriate screwdriver or other implement to twist this potentiometer to see if it improves your readings.

MAKE YOUR ROBOT FOLLOW A LINE AUTONOMOUSLY

When your sensor is successfully detecting the line, it's time to give your robot the ability to follow the track.

By the end of this project, you'll have connected up a second sensor and programmed the logic behind a fully autonomous program, giving you a line-following robot.

Wiring Up the Second TCRT5000 Line-Following Sensor Module

You've already wired up the first TCRT5000 module, so now you only need to wire up the second one. If you don't have the first one connected, flip back a few pages and follow the wiring guide in the previous project.

The process for connecting your second sensor is as follows:

1. Take a F-F jumper wire and connect the VCC pin of your second TCRT5000 module to physical pin 17 on your Raspberry Pi. This is another +3.3 V pin and will power the sensor.

2. Now, use a jumper wire to connect the GND pin of the new module to the common ground rail on your breadboard.

3. Use a wire to connect the data output pin of your second sensor (OUT) to physical pin 23 on your Pi. This is BCM 11. The wiring for the second sensor, with the first sensor omitted, should look like Figure 7-13.

FIGURE 7-13

Second line-following sensor connected to power, ground, and data

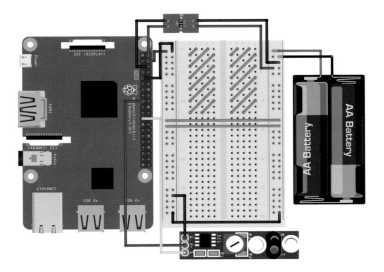

With both line-following sensors connected, your breadboard should look something like Figure 7-14.

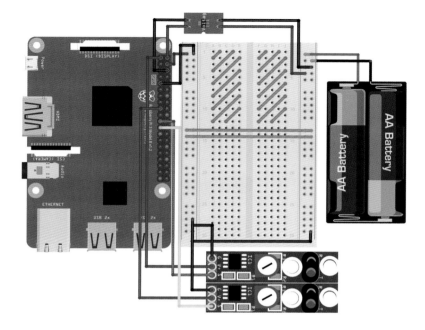

FIGURE 7-14

Both line-following sensors connected to the Raspberry Pi

Mounting Your Sensors

Now you need to mount your TCRT5000 modules onto the underside of the front of your robot.

To avoid having to change the code demos later, make sure that the TCRT5000 module you wired up *first* (the one connected to BCM 9) is on the *left side* of your robot, and the *second* TCRT5000 module (connected to BCM 11) is on the *right side* of your robot.

You can mount these sensors to your chassis in whatever way you like. If you're using a LEGO-based robot like mine, I recommend attaching two small stalks made of 2×2 LEGO blocks to the bottom of your chassis; mine are roughly four blocks deep. You can then also use sticky tack to affix your modules onto the bottom of the stalks as I have done (see Figure 7-15). I have routed the wires for the sensors through the gap in the middle of my LEGO chassis.

FIGURE 7-15

My IR sensors,
mounted on LEGO
stalks on the underside
of my robot

When mounting the modules, bear in mind that the optimal range from the optical sensor to the ground is between 1 mm and 8 mm. Also remember that the wider the gap between them, the more your robot will stray left or right before it corrects itself. For reference, mine are only an inch or two apart from each other and placed on either side of the front stabilizer.

Programming Your Robot to Follow a Line

With both your sensors wired up and mounted in place, it's time to write the code that will allow your robot to follow a line.

Open up a new program using Nano and call it *line_follower.py* as follows:

```
pi@raspberrypi:~/robot $ nano line_follower.py
```

The code in Listing 7-2 takes the theory and process behind line following we discussed previously and puts it all into practice. Take a moment to look over the code before you move on to the explanation.

```
     import gpiozero

❶   SPEED = 0.25

     robot = gpiozero.Robot(left=(17,18), right=(27,22))

❷   left = gpiozero.DigitalInputDevice(9)
     right = gpiozero.DigitalInputDevice(11)

     while True:
❸       if (left.is_active == True) and (right.is_active == True):
             robot.forward(SPEED)
❹       elif (left.is_active == False) and (right.is_active == True):
             robot.right(SPEED)
❺       elif (left.is_active == True) and (right.is_active == False):
             robot.left(SPEED)
❻       else:
             robot.stop()
```

LISTING 7-2

Following a line

The code here follows the same logical layout as previous projects in this book. We import gpiozero, then initiate a constant called SPEED and set it to 0.25 ❶. This value, which represents the speed of your robot throughout the program, can be set to anything between 0 and 1.

When you run this code for the first time, you'll discover that the speed at which your robot is traveling has a huge influence on its line-following ability. By defining a constant at the start of the program, you can easily tweak this setting without rifling through all of the rest of your code later.

At ❷ we set up the first TCRT5000 sensor as a digital input and assign it to the variable left. Then we repeat the process for the second line-following sensor, assigning it to the variable right.

We then begin an infinite while True loop that contains the main logic of the program: a series of if statements. This is where the line-following theory is implemented.

The code at ❸ deals with the scenario in which both sensors are active and therefore both reading white. With no line detected, the program assumes that the line is directly underneath and between the sensors, and the robot proceeds forward at its given speed.

Next, an elif statement ❹ catches the case when the left sensor is off and thus detecting black, but the right sensor is active and thus detecting white. In this scenario, we need to execute a corrective maneuver to turn the robot right.

Then, an almost identical `elif` statement ❺ accounts for when the right sensor detects the black line. The corrective maneuver turns the robot left this time.

Finally, at ❻ we use an `else` statement to deal with the only option left: neither sensor is active and therefore both are reading black. The exact behavior of your robot after this is up to you to define. In my code I have decided to stop its movement.

Running Your Program: Make Your Robot Follow a Line!

With your line-following program now complete, connect your robot to battery power and place it on the test track you created earlier. For the best results, ensure that the black line is directly underneath your robot and between the two TCRT5000 sensors (see Figure 7-16).

FIGURE 7-16

My robot in place, ready to follow the line

NOTE

You may find that you need to stick your track down to the floor so that the wheels of your robot don't move it around. If this is the case, just use some tape to securely and nonpermanently fix it in place.

Run your program with this command:

```
pi@raspberrypi:~/robot $ python3 line_follower.py
```

Your robot should now be moving around your track completely autonomously and sticking to the black line with no trouble at all. Admire your work as the robot scuttles around in an endless loop!

EXPERIMENTING WITH LINE FOLLOWING

Line following is a classic lesson in robotics, but as you can probably tell, it can take a serious amount of tweaking and hacking to improve the effectiveness and results. You'll probably have already noticed that you can alter many different factors to change the performance of your line follower. Here are some of my suggestions.

Change the Track

While a simple loop is a great first-time test bed, it quickly becomes boring. To challenge your robot, try creating a more adventurous track. Use tighter corners, longer straights, and more advanced geometries! You can even get inspiration for new layouts from real race tracks—for example, the Monaco Grand Prix circuit in Figure 7-17.

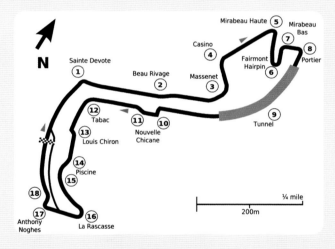

FIGURE 7-17

The Monaco Grand Prix circuit represented as a black line that your robot could follow

Change the Speed

The faster your robot is moving around the track, the less time it has to react to the information that the TCRT5000 sensors provide. At faster speeds, you may find that your robot drifts away from the line. You can change its speed by going back into the program and editing the SPEED constant defined at the top of the code.

(continued)

You may also find that you want to change the speed of your line follower for different movements. For example, your robot may perform better if it goes around corners slower. You could add this functionality into the program by creating another constant called `CORNER_SPEED` and using it in the functions for turning left and right.

As you fine-tune the speed of your line-following robot, time it as it goes around the track. See how fast you can get it to accurately follow the line. You could even race against some friends and have a line-following racing tournament to see whose code is best.

Change the Position of Your Sensors

The farther apart the IR sensors are, the more your robot will stray from the line before executing a corrective maneuver. This could potentially lead to your robot fish-tailing along the line in a zigzag pattern. Experiment with moving the modules closer together to see how this affects how well your line follower operates.

You should find that the closer together the modules are, the more the robot will turn left and right to get back on track. This will mean that your robot follows the line more accurately, but at what cost? Does it affect the speed at which it can complete the course? Play around and find out!

Add More Sensors

If you want to take this even further, add more line-following sensors. Just like with obstacle avoidance, the more information your robot has about its position, the more intelligent the program behind it can be. If you can add a third, fourth, or even fifth TCRT5000 module, then you can judge the degree by which your robot has strayed from the line. You could then use this information to change the magnitude of course corrections. For example, if the robot is far away from the line, it could execute a faster corrective maneuver.

SUMMARY

In this chapter you've given your robot the ability to autonomously follow a line. We have covered everything from the theory of line following to the sensors you need and the code behind them.

In the next chapter I'll show you how to use the official Raspberry Pi Camera Module to recognize and follow a colored ball!

8
COMPUTER VISION: FOLLOW A COLORED BALL

AS HUMANS, WE USE OUR EYES AND BRAINS TO SEE AND COMPREHEND THE WORLD AROUND US. ALTHOUGH THIS HAPPENS AUTOMATICALLY FOR US, VISION IS ACTUALLY AN IMMENSELY COMPLICATED SERIES OF PROCESSES.

Computer vision is an advanced field of computer science and engineering that aims to enable computers and machines to see and understand their surroundings at least as well as humans, if not better. In this chapter you'll learn some principles of computer vision, and then I'll show you how to use a camera to enable your robot to recognize and follow a colored ball.

THE COMPUTER VISION PROCESS

Consider the process behind seeing, recognizing, and reacting to a colorful item. First, the image of that item passes through your eye and strikes your retina. The retina does some elementary analysis and then converts the received light into neural signals, which are sent to your brain and analyzed thoroughly by your visual cortex. Your brain then identifies the item and gives instructions to your muscles.

What is remarkable is that this all happens in a fraction of a second and with no conscious effort.

Even with that simplified explanation, you can appreciate the complexity of vision. Getting computers to complete a similar series of tasks is something that people in the field of computer vision have worked on tirelessly for *decades*.

There are three distinct tasks any computer vision system must be able to do:

See   Biological beings generally see through their eyes. Computers must use their digital equivalent: cameras. Cameras work by using a lens to focus light onto a digital sensor. This sensor then converts the light into digital information: an image or frame of a video.

Process   After the input has been captured, it must be processed to extract information, recognize patterns, and manipulate data. In nature, this is the role of the brain. For computer vision, it is the role of code and algorithms.

Understand   The information must then be understood. The computer may have detected and processed a pattern, but what is the pattern and what does it mean? Again, this important step relies on code and algorithms.

When these three elements work together, the computer can handle all sorts of vision-based problems, including the one we're going to tackle in this chapter: we'll give your robot the ability to detect, recognize, and move toward a colored ball that might be anywhere in its surrounding environment.

THE PARTS LIST

You'll need two new items for this next project:

- A colored ball
- Standard Pi Camera Module

Initially, you'll also need access to another computer to remotely view images taken by the Pi Camera Module during the configuration stage of this project. This can be the computer you're using to SSH into your Pi. If you haven't been using SSH over the course of this book, then you can just hook up your Pi to an HDMI display during the configuration.

Let's take a closer look at the new parts.

The Target: A Colored Ball

First you'll need a colored ball to act as the target your robot will seek out and follow. Your ball should be a bright color that doesn't appear much elsewhere in the room, to help your robot differentiate it from other objects. I'd recommend something distinctive and not too large, like the balls shown in Figure 8-1. The one I'm using is a bright yellow stress ball roughly 2 inches in diameter. You probably already have something suitable lying around your house, but if not, you should be able to pick up a similar one online for a few dollars.

FIGURE 8-1

Example colored balls for targets; I'm using the yellow one on the left.

The Official Raspberry Pi Camera Module

To give your robot the ability to see, you will need a camera. In this project, we'll be using the official Raspberry Pi Camera Module, shown in Figure 8-2.

FIGURE 8-2

The official Raspberry
Pi Camera Module

The Camera Module is a Raspberry Pi add-on board designed and produced by the Raspberry Pi Foundation. The latest model features an 8-megapixel sensor and is less than 1 inch square in size! Not only does it take great still photos, but the Camera Module is also able to shoot full-HD 1080p video at 30 frames per second. If you have the older 5-megapixel model, don't worry: it is fully compatible with this project and used in exactly the same way.

The 6-inch ribbon cable on the Camera Module attaches to the CSI (Camera Serial Interface) port on the Raspberry Pi, shown in Figure 8-3. It is compatible with all models, including the latest version of the Pi Zero.

FIGURE 8-3

The CSI interface port
on the Raspberry Pi

You can buy the official Raspberry Pi Camera Module online at the usual retailers. It costs approximately $30.

When looking online you may notice that there are actually two different official Camera Modules: the normal camera and the *NoIR* version, which can be used for night vision. You need the *standard* Camera Module. You can easily tell the two boards apart by their color difference: the circuit board of the normal Camera Module is green, whereas the NoIR is black.

CONNECTING AND SETTING UP YOUR CAMERA MODULE

Before you attach the Camera Module, ensure that your Pi is switched off. Then follow this process:

1. Locate the CSI port on your Raspberry Pi. For all the full-size models of Raspberry Pi, this is found between the HDMI port and the 3.5 mm audio jack, and is helpfully labeled CAMERA.

2. Next, open up the port by gently but firmly grabbing it from either side and pulling it upward (see Figure 8-4). This can be a delicate operation, and it often helps to get your fingernails underneath the sides.

3. Insert the ribbon cable on your Camera Module all the way into the CSI port with the silver contacts facing *away* from the 3.5 mm audio jack and Ethernet port (see Figure 8-5). This orientation is critical: if you insert the Module's cable the other way around, it won't be connected properly and you will not be able to use it!

NOTE

The Pi Zero has a mini-CSI connector rather than the full-size one found on all other Raspberry Pi models. To connect a Camera Module to the Zero, you'll need to also purchase a mini-CSI-to-CSI cable. These are branded online as "Pi Zero Camera Cables" and cost around $5. Keep in mind that this project requires intense image processing and code that will run better on the faster full-size Raspberry Pi models than it will on the Zero or original Raspberry Pi models.

FIGURE 8-4

The CSI port opened for cable insertion

4. Then, while holding the ribbon cable in place, place a finger on both sides of the CSI port and push it back down at the same time. If both sides don't close simultaneously, then one side will not close properly and the cable may come loose. Figure 8-6 shows a properly attached ribbon cable. Notice that a fraction of the silver contacts are just visible, and they're all parallel with the board.

5. Finally, to ensure that the Camera Module is connected properly, give the ribbon cable a gentle tug near the CSI port. It should stay rigidly in place. If the cable comes detached or slips, don't worry—just remove it and repeat these steps.

If you want to connect the Camera Module to a Raspberry Pi Zero, the process is similar. Find the mini-CSI port on the right-hand side of the board and open it using a finger on either side of it. Then, make sure the silver contacts face downward toward the board when you insert the Pi Zero camera cable. See Figure 8-7.

Mounting Your Camera

Now that you have the camera connected to the Pi on your robot, you need to mount it in an appropriate position. I recommend using some sticky tack to affix it to the front of your robot, relatively low to ensure it will have a clear view. To do this, I've used a 2 × 2 LEGO brick to create some mounting space (see Figure 8-8). Also make sure that your camera is positioned the correct way up, like mine is in the picture.

The Camera Module is quite delicate, so handle it with caution. Try not to contort the cable too much and make sure to not leave any kinks in it. If at any point the ribbon cable comes loose from the Raspberry Pi, just reattach it the same way as before. If it comes loose from the connector on the actual module then you can also just reattach it. This is done in the same way: use your fingers to lever open the module's CSI port and then insert the cable with the silver contacts facing down and toward the PCB.

Enabling the Camera and VNC, and Setting the Screen Resolution

To use the camera in Raspbian, you first need to enable it. If you followed all of the instructions in Chapter 1, you'll already have done part of this. On top of that, we'll need to enable VNC for this project and manually set up the right screen resolution. Here's the full process.

To do this setup, we'll use the configuration tool, `raspi-config`. Open the command line and enter the following:

```
pi@raspberrypi:~/robot $ sudo raspi-config
```

You should see the same blue configuration screen you met when configuring the audio output of your Raspberry Pi a few chapters ago. Using the arrow keys, scroll down to **Interfacing Options** and then press ENTER. This opens up a new menu shown in Figure 8-9.

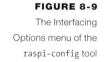

FIGURE 8-9

The Interfacing Options menu of the raspi-config tool

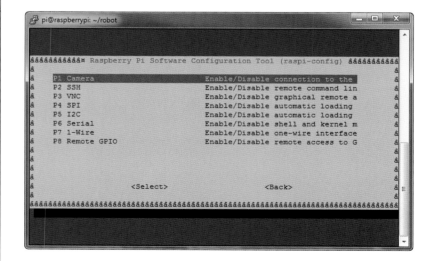

Select **Camera** by pressing ENTER again. Then, when asked whether you'd like to enable the camera interface, use the left/right arrow keys to select **Yes** (see Figure 8-10).

FIGURE 8-10

Enabling the camera using
raspi-config

You'll see a message confirming that the camera interface has been enabled and then you'll return to the original menu.

While you are in the raspi-config tool, it is also important to ensure that VNC is enabled. VNC will be fully explained in the next section, but for now just scroll down to **Interfacing Options** again, and then select **VNC** (see Figure 8-11). Press ENTER to enable it.

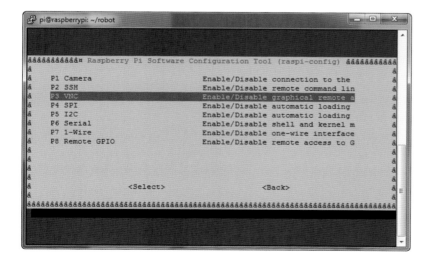

FIGURE 8-11

Selecting VNC from
Interfacing Options

You'll be sent back to the original menu. Before you can exit the configuration tool, you have to do one last thing: manually set the screen resolution of your Pi. This will ensure that when we use VNC later, the screen will be set up in the right way. To set the resolution, from the original menu, scroll down to **Advanced Options**, and then scroll down and select **Resolution** (see Figure 8-12).

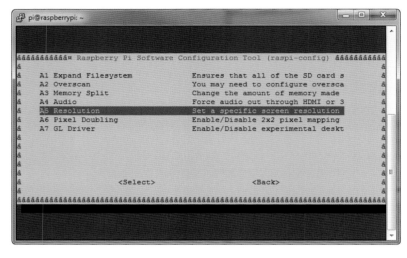

You will then be prompted to choose a screen resolution. Use the arrow keys to scroll down to the full HD option. It will look something like "DMT Mode 82 1920x1080 60Hz 16:9." Press ENTER on this option, and your screen resolution should now be set! You'll be returned to the original menu.

Exit the configuration tool by pressing the right arrow key twice (highlighting **Finish**) and then pressing ENTER. Reboot your Raspberry Pi if prompted to do so.

TAKING A TEST PHOTO

With your Camera Module connected and enabled, let's test it by taking a photo. This is easy to do from a remote terminal with a simple command, but by using that method you won't be able to view the image it took in the text-based environment! This is where VNC, the option you enabled previously, comes in.

Controlling Your Pi's Desktop Remotely with VNC

VNC stands for *Virtual Network Computing*. It allows you to remotely view and control your Raspberry Pi's desktop from another computer, a little bit like what you've been doing with SSH, but for the full graphical user interface (GUI) rather than just the terminal. Since we can view the Pi's GUI using VNC, you'll easily be able to see any photos that you take with the Raspberry Pi Camera Module using Raspbian's built-in image viewer.

Installing and Making a Connection with VNC Viewer

You have everything set up on the Raspberry Pi end, and now you need to download a VNC viewer on the computer you want to view the images on. We'll use some free software called *VNC Viewer* from RealVNC, which is compatible with Windows, Mac, Linux, and more. To install the software, follow this process on your machine:

1. On your computer, open a web browser and go to *https:// www.realvnc.com/en/connect/download/viewer/*. You should see the download page for the VNC Viewer software, shown in Figure 8-13.

2. From here, select your operating system and click the **Download** button. Once the software has downloaded, go through the installation wizard and agree to the terms of service. A few minutes later, everything should be installed and ready!

NOTE

If you've been following along with this book and not using your Raspberry Pi wirelessly over SSH, don't worry! You can still follow the steps after this section and view the results with your Pi connected to a monitor over HDMI.

FIGURE 8-13

The VNC Viewer software download page

With VNC Viewer now installed, run it. You should see a window with a box at the top (Figure 8-14); here, you'll enter the IP address of your Raspberry Pi, which you should already know since you've been using it to connect to your Pi over SSH.

An authentication box will appear asking you for a username and password. Enter the login details of your Raspberry Pi and click **OK**. If you haven't changed the default user, then the username will be `pi` and the password will be whatever you set it to in Chapter 1.

FIGURE 8-14

Setting up VNC Viewer
to connect to my
Raspberry Pi

A new window will appear displaying your Pi's desktop (see Figure 8-15). Here you can access and use everything just the same as if your Pi were plugged into an HDMI monitor.

FIGURE 8-15

A terminal and file
manager in my
Raspberry Pi desktop
environment, viewed
with VNC

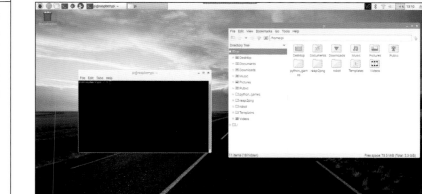

Taking and Viewing a Photo Using the Raspberry Pi Camera Module

Now that you have everything set up, you can take your test photo! We'll use a built-in command-line tool called `raspistill`. Open up a terminal, either through an SSH connection or by using a terminal window in the desktop environment over your VNC connection, and enter the following command to take a photo:

```
pi@raspberrypi:~ $ raspistill -o test.jpg
```

After a 5-second delay (to make sure you can get in front of your camera or frame your shot), this command will complete. If you see no output, that's great news! There is no success message for this particular command. This instruction takes a picture and saves it as *test.jpg* in the directory where the command was run—in this case, the default home directory.

To view the image in the VNC desktop, click the **File Manager** icon in the VNC desktop environment (it looks like a collection of folders, as shown in Figure 8-16).

CAMERA TROUBLESHOOTING

If running the `raspistill` command gives you a scary error similar to the following, don't worry!

```
pi@raspberrypi:~ $ raspistill -o test.jpg
mmal: mmal_vc_component_enable: failed to enable component:
ENOSPC
mmal: camera component couldn't be enabled
mmal: main: Failed to create camera component
mmal: Failed to run camera app. Please check for firmware
updates
```

You most likely just haven't connected your camera properly. If you get this error message, or any other, check the ribbon cable connections between the Camera Module and your Pi. Also ensure that you have enabled the camera interface properly—turn back to page 172 for guidance.

FIGURE 8-16

The File Manager icon

Navigate to the directory you ran the `raspistill` command from (I ran mine in the default home directory), locate *test.jpg*, and double-click it. You should see the photo that you just took in the Image Viewer (see Figure 8-17).

FIGURE 8-17

The *test.jpg* image
in Raspbian's Image
Viewer over VNC (top);
the *test.jpg* image,
which features a stapler,
Carl Sagan's *Cosmos*,
and two colored balls
for use later (bottom)

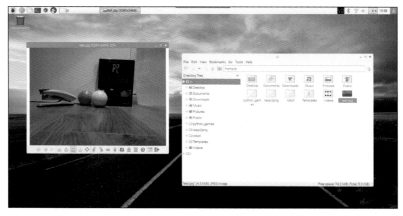

MAKE YOUR ROBOT SEEK AND FOLLOW A BALL

Now you have your camera hooked up and a successful image test under your belt, it's time to move on to the advanced project in this chapter: making your robot recognize and follow a colored ball. But first: a quick lesson in some important theory.

Understanding the Theory Behind Colored-Object Recognition

How do we get a robot incapable of independent thought as we know it to detect and identify a particular object?

As your robot moves around, the position of the ball relative to your robot will be constantly changing, so the first thing we need is a continually refreshing view of what is in front of your robot. The Camera Module provides this view through a stream of images, often referred to as *video frames* or just *frames*.

Each image, like the one in Figure 8-18, will need to be analyzed to identify whether or not it contains your colored ball. To do this, we will apply various image processing techniques.

FIGURE 8-18

An untouched image from the Camera Module, ready to be analyzed

The first step is to convert the image, which is in an RGB format, into an *HSV* format. We discussed RGB in Chapter 6, but I'll briefly summarize here. *RGB* stands for red, green, and blue. Each pixel of the image from the Camera Module in Figure 8-18 is made out of a combination of these three colors, represented as three numbers between 0 and 255—for example, [100,200,150].

Computers use RGB for *displaying* colors, but for processing images and the color data they contain, the HSV color format is much more appropriate. *HSV* stands for hue, saturation, and value, and it is just another way of digitally representing color with three parameters. HSV is a bit more complicated to understand and represent than RGB, but it is often easiest to get your head around when viewed as a cylinder (see Figure 8-19).

FIGURE 8-19

An HSV cylinder

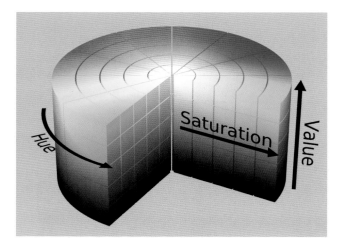

Hue is the color portion of the HSV model, and it is expressed as a number from 0 to 360 degrees. Different sections of this range represent the different colors (see Table 8-1).

TABLE 8-1

The Hue Ranges of HSV

COLOR	ANGLE
Red	0–60
Yellow	60–120
Green	120–180
Cyan	180–240
Blue	240–300
Magenta	300–360

Saturation is the level of white in a color, from 0 to 100%. *Value* works alongside saturation and can be thought of as brightness; it describes the intensity of the color from 0 to 100%.

By converting each image into an HSV format, your Pi is able to separate just the color component (hue) for further analysis. This means that the computer should still be able to recognize a colored object, regardless of the environment and its lighting effects. This would be incredibly difficult to achieve in an RGB color space. Figure 8-20 shows an RGB implementation of the HSV data.

FIGURE 8-20

HSV data of the image

As you can see in Figure 8-20, the yellow ball my robot will follow is now clear and distinctive. There is no possible way that it could be confused with the red ball behind it, or any of the other objects in the frame. Remember, though, that this is an RGB *implementation* of HSV color. The hue value of a color is not something that we can see in the same way with our own eyes.

The next stage in the process is to look for and identify any color that matches the one we are searching for. In my case, I want to match all the parts of the image that are the same color as my yellow ball. This forms a *mask* (see Figure 8-21) that simply keeps the parts of the image we want and removes the parts we don't want. You can see that it keeps only the areas that contain my desired color, yellow.

FIGURE 8-21

Masking out the colors and areas of the image that are not the same shade of yellow as the ball. Notice that there are areas on the ball (bright reflections/shadowy regions) that aren't picked up!

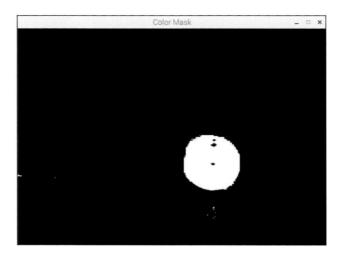

Now that the relevant colored objects have been isolated, the next step in the process is to identify the largest patch of color. Notice how there are other (albeit small) readings of yellow in Figure 8-21? If you don't program this right, your robot might get confused and head toward those instead of the desired ball. This could be disastrous—after all, you don't want it to get distracted by distant bananas!

Assuming that the largest part of the mask is the colored ball, the next stage is to actually find that largest area. We do so by drawing a contour (like an outline; see Figure 8-22) around each detected object in the mask. We can work out the area of each contour using some basic mathematics. The largest area is then identified and assumed to be our target ball!

FIGURE 8-22

Contour drawn around the largest single object in the mask

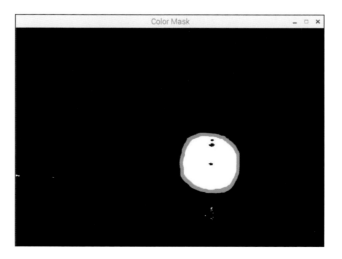

After this, we simply need to program the robot to move toward the object. If the target is to the right of the robot, move right. If it is to the left of the robot, move left. If it is in front of the robot, move forward.

And that's all there is to making a ball-following computer vision system with your Raspberry Pi! It's time to put it into practice.

Installing the Software

You'll need a couple of Python libraries to enable computer vision. Most notably, we will be using OpenCV, a free and open source library of programming functions for real-time computer vision. You will also need the PiCamera Python library to manipulate and handle the Camera Module in Python, though this is included by default in the latest versions of Raspbian.

To install the dependencies for the OpenCV Python 3 library, enter the following command into the terminal:

```
pi@raspberrypi:~ $ sudo apt-get install libblas-dev
libatlas-base-dev libjasper-dev libqtgui4 libqt4-test
```

When prompted as to whether you want to continue, press Y and then ENTER. This command will take a few minutes to execute.

Now, you can use `pip` (the Python software management tool we have used previously) to install OpenCV for Python 3. Enter the following command:

```
pi@raspberrypi:~ $ sudo pip3 install opencv-python
```

After the OpenCV installation has finished, check that you have the PiCamera library installed with the following command. It will most likely inform you that you already have the newest version, but if not, proceed with the installation:

```
pi@raspberrypi:~ $ sudo apt-get install python3-picamera
```

And that's all you will need!

Identifying the HSV Color of Your Colored Ball

To identify a ball of a specific color, your Raspberry Pi robot needs the HSV value of that color. The Pi will use this value to compare each part of each image to see whether it is the color of the ball that you want your robot to follow.

NOTE

If you get any errors during this installation process, or it just doesn't look like it has gone correctly, visit the book's website at https://nostarch.com/ raspirobots/ to check for any changes and get further guidance.

Your ball is probably a different color than mine, so you'll need to find out the exact HSV value for yours. Even if you have a yellow ball as I do, it may be a slightly different shade from mine!

There are various ways to identify the hue value you'll need, but I've found that the best method is to try out various values on your Raspberry Pi and observe the effects. The aim is to find the one that matches your particular colored ball. To help you do this, I have created a test program, *hsv_tester.py*, which you can find in the software bundled with this book (*https://nostarch.com/raspirobots/*). The next section will walk you through running the program.

Running the HSV Test Program

Place your robot in a well-lit environment, with your colored ball approximately a meter in front of it. Then boot up the Pi on your robot and view its desktop remotely over VNC. Next, open up a terminal application in the desktop, locate the *hsv_tester.py* program, and run it using the command:

```
pi@raspberrypi:~/robot $ python3 hsv_tester.py
```

You'll see a prompt asking you for a hue value between 10 and 245. Try approximating the hue value of your ball; the ranges in Table 8-1 should give you a rough idea of what end of the spectrum to start at. Mine is yellow, so I'm going to guess 40. When you enter the value, you'll see four new windows appear, showing you the different stages of image processing discussed previously (see Figure 8-23).

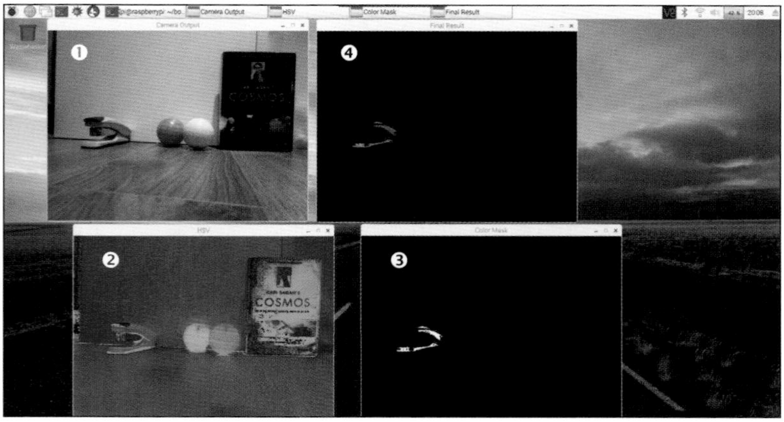

FIGURE 8-23

The four different windows of the HSV tester program

The first window, titled *Camera Output*, is a raw RGB video output directly from the Camera Module ❶. The second window, titled *HSV*, is the same video converted into HSV format ❷. Then, the window titled *Color Mask* displays the parts of the image that match the hue number you provided ❸. Finally, the window titled *Final Result* overlays the color mask with the original video feed, showing you the isolated area ❹.

If the mask in the Final Result window is more or less ball-shaped, you have your hue!

It is unlikely that you'll get your hue value correct the first time, just as I haven't—the hue didn't match my ball but instead it got parts of the stapler in the frame! To try again, select any of the output windows (but not the terminal window), and press Q on your keyboard. This will freeze the video output, and you can then go back to the terminal and enter another value to try.

Play around and tweak the hue until you get a definitive match for your colored ball. After a little while, I found that my magic number was 28. When you have found yours, you should see that the majority of your ball and not much else is left in the frame, like mine in Figure 8-24.

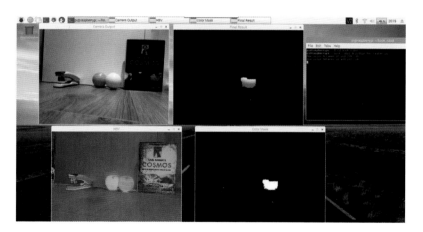

FIGURE 8-24

My output windows after I correctly identified my hue number as around 28

Make a note of the value, as you'll need it soon. After you've found the right number, close the HSV tester program by pressing CTRL-C in the terminal window.

Programming Your Raspberry Pi to Follow a Ball

With all of the groundwork in place, you can now program your Raspberry Pi robot to follow a ball! The program we'll be using is relatively long and more advanced than anything you've met so far, so I recommend downloading it from the software bundle rather than copying it from this book to minimize typos. The program is called *ball_follower.py* and you can check it out with the command:

```
pi@raspberrypi:~/robot $ nano ball_follower.py
```

This program is 75 lines long, so for the explanation I've split it into sections and will explain how each part works. If you're more interested in just running the code first and understanding how it works later, skip ahead to the next section on page 193.

Importing Packages and Setting Up the Camera Module

First we'll import the packages we need and set a few things up, as shown in Listing 8-1.

❶ ```
from picamera.array import PiRGBArray
from picamera import PiCamera
import cv2
import numpy as np
import gpiozero
```

❷ ```
camera = PiCamera()
```
❸ ```
image_width = 640
image_height = 480
```
❹ ```
camera.resolution = (image_width, image_height)
camera.framerate = 32
rawCapture = PiRGBArray(camera, size=(image_width, image_height))
```
❺ ```
center_image_x = image_width / 2
center_image_y = image_height / 2
```
❻ ```
minimum_area = 250
maximum_area = 100000
```

The first lines of the program ❶ import the necessary libraries, including various parts of the PiCamera library needed to allow us to use the Camera Module in Python. We also import the OpenCV library, cv2; gpiozero as usual; and the NumPy library, np. NumPy is a Python package used for scientific computing and will be useful for manipulating the image data later.

At ❷, we initiate a PiCamera object and assign it to the camera variable for use throughout the program. Next we define the size ❸ and resolution ❹ of the images being fed from the camera. We won't need full-HD video frames and they would just slow down speed and performance, so we downgrade the resolution to standard definition: 640×480 pixels. On the lines following this, we specify the frame rate of the camera and the raw capture setup.

The two lines at ❺ work out where the center of the image is. This information will be used later to determine and compare where the ball is in the frame.

Then at ❻, we set a minimum and maximum area for the colored ball. This prevents your robot from detecting and following any colored object smaller than 250 square pixels or larger than 100,000 square pixels. These are arbitrary numbers that I've found work pretty well, but if you want to change them later, feel free!

Setting Up the Robot and Color Values

This section deals with the final part of the setup process, shown in Listing 8-2.

```
   robot = gpiozero.Robot(left=(17,18), right=(27,22))
❶ forward_speed = 0.3
   turn_speed = 0.2

❷ HUE_VAL = 28

❸ lower_color = np.array([HUE_VAL-10,100,100])
   upper_color = np.array([HUE_VAL+10,255,255])
```

We set up the robot and its motor pins as before and then define two variables for both the forward and turning speeds ❶, with the values 0.3 and 0.2, respectively. This will limit the speed of your robot when it moves toward your colored ball. Again, these are arbitrary numbers, so you can change them if you find that higher or lower values work better for you and your robot.

At ❷, we set the number for the hue. This is a value you *must change* to the value you found earlier using the HSV tester program. I have set mine to 28.

Next we set a range of values for the robot to check for instead of a precise one ❸. That way, changes in the environment, like the room's lighting and how bright it is, will still fall within this small range and therefore the ball will continue to be detected. We do this by using arrays to create the upper and lower bounds of the color in HSV format.

In programming, an *array* is a collection of information where each piece of data in the array has an *index*, or location, associated with it. Arrays can be as long as you like and can store anything you want, from people's names to types of animal to lists of numbers. In Python, the first piece of data in an array has an index of 0, the second piece has an index of 1, the third piece has an index of 2, and so on—in other words, Python starts counting the items in an array at 0. This means that, in a Python program, you could ask for the piece of information stored in the array at index 3, for example, and it would return the data found in the fourth position.

In this situation, we use arrays to represent the HSV format, as each HSV color can be described with three numbers (the hue, saturation, and value). Notice that we actually search for hues ±10 in either direction, and that the range for the saturation and value components of the color goes from 100 to 255. This ensures that the robot will look for a broader range of colors in each frame from the camera and improves the odds that it will detect the target colored ball.

These arrays are available to use through the NumPy library we imported. We use NumPy here because it is a highly optimized library for fast array calculations. This gives us the speed necessary to access and analyze each pixel of each frame.

Analyzing the Camera Frames

The third section of the program is shown in Listing 8-3. This is where the bulk of the code and computer vision process starts.

LISTING 8-3

Starting capture for loop, converting image, and finding contours

```
❶ for frame in camera.capture_continuous(rawCapture, format="bgr", ↵
  use_video_port=True):
  ❷ image = frame.array

  ❸ hsv = cv2.cvtColor(image, cv2.COLOR_BGR2HSV)

  ❹ color_mask = cv2.inRange(hsv, lower_color, upper_color)

  ❺ image2, countours, hierarchy = cv2.findContours(color_mask,
     cv2.RETR_LIST, cv2.CHAIN_APPROX_SIMPLE)
```

At ❶, we initiate a for loop that, translated into plain English, reads: "for each frame from the Camera Module, do this."

Next, the information from the current frame is saved to the variable image as an array ❷. The RGB data of image is then converted into an HSV format ❸ using an OpenCV function, cvtColor().

Once the HSV data has been acquired, the color mask, which keeps *only* your desired color, is created ❹. We use OpenCV's inRange() function so that the mask keeps all of the colors that fall between the lower and upper bounds of your color choice.

The next stage of the process is to draw lines around each distinct object in the mask so that the area of each detected object can be compared later. We do this at ❺, using OpenCV's findContours() function.

When you have everything set up, run the program with this command:

```
pi@raspberrypi:~/robot $ python3 ball_follower.py
```

Your robot should spring to life and start hunting down your colored ball. Play a game of fetch with your new smart pet!

As usual, press CTRL-C to stop this program.

Experimenting with Image Processing

As with the line-following project in the previous chapter, computer vision and image processing are areas of computer science and robotics that lend themselves to fine-tuning in order to improve results and capabilities. Here are some suggestions for you to play around with.

Color and Object

While your colored ball is a great starting target for your robot, you can easily take it even further. For example, why not introduce a second color by scanning each frame for a secondary set of HSV values? Make your robot follow both yellow and red objects, for example. Remember that you can go back to the HSV tester program to work out the hue and color codes of other shades!

You're also not just restricted to balls. You can make your robot follow or seek anything that is primarily a single color. Experiment with other objects that you have lying around!

Speed

How fast your robot moves does have a big effect on the quality of its image processing: usually, the faster it goes, the more likely it is to miss your colored object. That said, feel free to play around with the speed values defined at the start of the ball-following program— you may be able to fine-tune and improve your robot's performance!

Minimum/Maximum Area of Object

Experiment with the minimum and maximum area of the target object. Remember that your robot by default will not move toward anything smaller than 250 square pixels, and it will stop at anything larger than 100,000 square pixels.

(continued)

By changing these numbers, you can make your robot move toward targets that are potentially smaller or even stop closer to the target. A fun idea is to increase the maximum area to a point where your robot won't stop when it gets close to your colored ball. The result is that your robot usually ends up bumping into the ball and "kicking" it . . . only to trundle after it again and repeat the process!

Remember that each frame of the video feed from your Camera Module measures 640×480 pixels, so a value of 307,200 is the maximum number of square pixels possible.

Avoidance Behavior

At the moment, your robot loves your colored ball, but what if you made it so the opposite was true? Try editing the program so that your robot runs away from the object, rather than toward it.

An extension of this would be to make your robot move toward certain colored balls, but run away from others. For example, it could love red balls but be terrified of yellow ones!

SUMMARY

In this chapter, you've given your robot the advanced ability to seek out, recognize, and follow a colored ball. You have learned the basics of image processing and implemented an entire computer vision process in Python using the official Raspberry Pi Camera Module.

And with that you have completed the projects section of this book! Your little robot is now all grown up, and you're its proud parent. This isn't the end of the line though; check out "Next Steps" on page 195 for some guidance and suggestions for continuing your adventures in robotics, programming, and Raspberry Pi.

NEXT STEPS

NOW THAT YOU'VE COME TO THE END OF THIS BOOK, IT'S TIME TO GO OUT INTO THE WORLD AND MAKE YOUR OWN ROBOTS AND OTHER RASPBERRY PI PROJECTS! I HOPE YOU'VE LEARNED AND DEVELOPED THE SKILLS NECESSARY TO HELP YOU ON YOUR FUTURE COMPUTER SCIENCE ADVENTURES.

But know that you're not alone! There are countless resources online and offline that can help take you further in your next steps. Here are some suggestions.

THE RASPBERRY PI GUY

I run the popular Raspberry Pi YouTube channel *The Raspberry Pi Guy*, where I provide free tutorials and educational videos. Here I have videos that cover all manner of Raspberry Pi–based subjects—from basic robotics to DIY electric skateboarding—in an easy-to-follow and accessible way (see Figure A-1).

FIGURE A-1

The Raspberry Pi Guy YouTube channel

You can take a look using the following links, and don't forget to subscribe if you like what you see.

The Raspberry Pi Guy YouTube channel
(*https://www.youtube.com/TheRaspberryPiGuy/*)

The Raspberry Pi Guy website
(*https://www.theraspberrypiguy.com/*)

GET IN TOUCH!

If you've enjoyed this book and want to reach out or share your progress, I'd absolutely love to hear from you via Twitter. You can follow and tweet to me *@RaspberryPiGuy1*. I'll be sure to see your tweets and reply, especially if you use the hashtag #raspirobots.

OTHER WEBSITES

The Raspberry Pi and the field of robotics both have massive, open, and incredibly welcoming online communities that you can read and contribute to. Here are just a few websites that you can learn and get inspiration from:

The official Raspberry Pi Foundation website (*https://www .raspberrypi.org/*) This is the site you used to download your Raspberry Pi's operating system at the start of the book. The Foundation's website has a plethora of educational resources and projects for all levels and abilities. You'll also find the Raspberry Pi forums, where you can sign up for an account and interact with other people with similar interests, problems, and questions; this is the perfect place to ask for help if you get stuck with a future project. There is even a dedicated section for "Automation, sensing, and robotics"! The site also includes a regularly updated blog featuring the latest news from the community.

Adafruit Learning System (*https://learn.adafruit.com/*)
This is a collection of resources and online lessons from the electronics community surrounding Adafruit. Here you'll find detailed tutorials for a huge variety of hardware and software, all with open source code and help. This is a good place to go if you need inspiration for your next project!

The official Python website (*https://www.python.org/*)
While this book has given you a great introduction to coding and the Python programming language, if you want to continue to develop your programming skills, take a look at the official Python website. You'll find all the documentation and guides you could possibly hope for! Further down the line, you may wish to explore other programming languages, such as C++ or Java, in which case simply search for tutorials on Google—there are quite literally millions!

New Atlas robotics news (*https://newatlas.com/robotics/*)
If you're interested in the latest news about robotics and progress in the field, check out the robotics section from New Atlas, an online news publication that covers technology. You'll find some really amazing stuff that might just inspire you to make something awesome and advance the field of robotics even further!

CLUBS AND EVENTS

There's a lot you can do online to learn about robotics and Raspberry Pi, but nothing can replace face-to-face clubs, events, and meetups. Fortunately, you'll find lots of opportunities for these in the computer science world. Here are some suggestions, but your location will affect what's available to you, so make sure you find out what's going on in your local area for clubs and events relevant to you!

Raspberry Jam (*https://www.raspberrypi.org/jam/*)
Raspberry Jams are independently organized community events for people of all ages to chat, learn, and share projects based around the Raspberry Pi. Jams happen in every corner of the globe and are run by all sorts of people. Raspberry Jams often include workshops for beginners, show-and-tell, and talks given by members of the community. People from all backgrounds and abilities, from beginner to professional, are welcome at these events. Use the *Find a Jam Near You* tool on the website to locate one in your area. If you ever come to one of the Jams in Cambridge, UK, or an event in Edinburgh, UK, you'll most likely find me there—please say hi!

Pi Wars (*https://piwars.org/*)   Pi Wars is an awesome challenge-based robotics competition for Raspberry Pi–controlled robots. Teams, both professional and amateur, build robots and then compete against each other in nondestructive tasks, such as obstacle courses, speed tests, and maze solving. It is usually held over a weekend and takes place in Cambridge, UK. You can go to the event as either a competitor or just a spectator; see the website for more information.

Code Club (*https://www.codeclubworld.org/*) and **Coder Dojo (*https://coderdojo.com/*)**   Code Club and Coder Dojo are worldwide networks of free, volunteer-led coding clubs for young people. Programmers and other people come together to help the younger generation learn how to program and develop their computer science skills. You can find out if there's one near you by using the tools on each website.

BOOKS AND PUBLICATIONS

If you've enjoyed learning about computer science and Raspberry Pi from this book, you may want to continue this journey with another book or publication. Here are some suggestions:

Python Crash Course, by Eric Matthes (No Starch Press, 2015)

20 Easy Raspberry Pi Projects, by Rui Santos and Sara Santos (No Starch Press, 2018)

Raspberry Pi User Guide, by Eben Upton and Gareth Halfacree (Wiley, 2016)

The MagPi (*https://www.raspberrypi.org/magpi/*), a free-to-read, monthly Raspberry Pi magazine by the Foundation that contains projects, coding, and other articles

RASPBERRY PI GPIO DIAGRAM

HERE YOU CAN FIND A FULLY ANNOTATED DIAGRAM OF THE GPIO PINS ON YOUR RASPBERRY PI. EACH PIN IS LABELED WITH ITS PHYSICAL NUMBER AND BCM NUMBER.

The physical numbers simply correspond to the pin's actual physical location, starting at 1 and going all the way down to 40. The BCM number on each pin (for example, BCM 25) is known as the *Broadcom* pin number, or the *GPIO* number. These numbers are used internally by your Pi's processor, and you usually need to use them inside GPIO Zero and other programming libraries.

Some pins have alternate functions specified in brackets; if you want to learn more about these, see the official documentation on the Raspberry Pi website (*https://www.raspberrypi.org/documentation/usage/gpio/*).

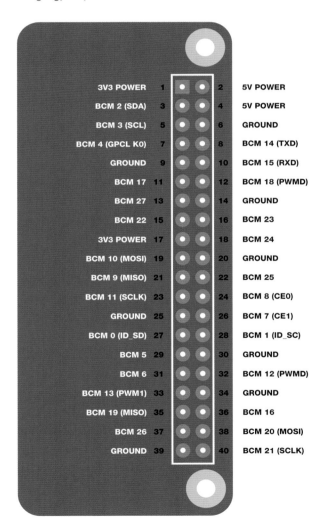

FIGURE B-1

The Raspberry Pi GPIO pins

RESISTOR GUIDE

RESISTORS ARE COMPONENTS SPECIALLY DESIGNED TO ADD RESISTANCE TO AN ELECTRICAL CIRCUIT IN ORDER TO REDUCE THE AMOUNT OF CURRENT PASSING THROUGH THE CIRCUIT. YOU'LL USE THEM AT VARIOUS POINTS IN THIS BOOK FOR A VARIETY OF TASKS, FROM ENSURING A SAFE CURRENT FOR AN LED TO CREATING A VOLTAGE DIVIDER CIRCUIT.

The exact resistance value of a resistor is measured in *ohms* (Ω). If you don't already know the value of your resistor, you can determine it using its colored bands. This table shows which value correlates to each color.

COLOR	FIRST BAND	SECOND BAND	THIRD BAND	MULTIPLIER	TOLERANCE
Black	0	0	0	1 Ω	
Brown	1	1	1	10 Ω	+/−1%
Red	2	2	2	100 Ω	+/−2%
Orange	3	3	3	1 KΩ	
Yellow	4	4	4	10 KΩ	
Green	5	5	5	100 KΩ	+/−0.5%
Blue	6	6	6	1 MΩ	+/−0.25%
Violet	7	7	7	10 MΩ	+/−0.10%
Gray	8	8	8		+/−0.05%
White	9	9	9		
Gold				0.1 Ω	+/−5%
Silver				0.01 Ω	+/−10%

A standard resistor has four colored bands. The first two bands represent the first two digits of the resistor's value. The third band is a multiplier that represents the number of zeros after the first two digits. The fourth band is the tolerance of the resistance; unless you're doing specialized and precise work, you most likely won't need to worry about the final band.

As an example, let's work out the value of the resistor shown in Figure C-1.

The first band is green, so its value is 5. The second band is blue, so its value is 6. The third band is yellow, which corresponds to four 0s. The fourth band is gold, meaning the tolerance is +/−5 percent.

The resistance value of this resistor is therefore 56 × 10,000 = 560,000 Ω, or 560 kΩ.

And that's all there is to working out a resistor's value from its colored bands!

NOTE

If you're color-blind like me, determining a resistor's value from its colored bands is almost impossible. I'd recommend keeping your resistors in organized and labeled bags so you can quickly and easily identify their value. If you're faced with an unknown resistor, you can use the resistance setting on your multimeter, to help you. Simply connect the two probes of your multimeter to either side of your resistor, and it will display its exact resistance.

FIGURE C-1

A four-band resistor

HOW TO SOLDER

SOLDERING IS THE PROCESS OF PERMANENTLY FUSING TOGETHER ELECTRONIC PARTS. IT INVOLVES MELTING A FILLER METAL ALLOY, CALLED *SOLDER*, BETWEEN TWO OR MORE COMPONENTS (SEE FIGURE D-1).

This process not only physically bonds the parts together, but also electrically connects them. Unlike using a breadboard, soldering is permanent. Over the course of your adventures in robotics, electronics, and Raspberry Pi, you will come across components that you need to solder (for example, wires onto motor terminals), so it is a key skill for you to learn and practice.

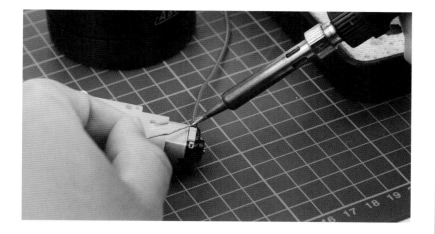

FIGURE D-1

Soldering a wire to a motor terminal—a common task in robotics

WHAT YOU NEED

In order to solder parts together, you'll need a few tools and materials:

- Soldering iron
- Solder
- Stand and tip cleaner

These should be available from the usual online retailers and local hardware stores. Let's take a closer look at each part.

Solder

First, you'll need the filler metal that you'll melt to create solder joints. Shown in Figure D-2, solder is a metal alloy with a relatively low melting temperature, usually between 180 and 200 degrees Celsius (356–392 degrees Fahrenheit).

In the past, lead was used in solder, due to its low melting point and fantastic electrical properties. However, we now know that lead is a heavy metal and toxic to humans. Consequently, lead-free solder, primarily made out of tin and copper, is the modern industry standard and what I recommend you purchase. If you're buying online, just search for "lead-free solder." It is usually provided in the form of a wire with a diameter between 0.5 and 0.8 mm.

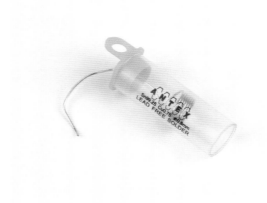

Soldering Iron

A soldering iron, shown in Figure D-3, is the tool that heats the solder to its melting point.

Soldering irons are available at a wide range of price points, generally starting as cheap as $10 and going higher than $100! As a beginner, you probably shouldn't get an expensive one with pro-level features. All you need is a decent iron that works reliably and doesn't give you any stress.

One common introductory soldering iron is the Antex XS25, which you can pick up for around $30. This is a quick-warming, fixed-temperature, quality tool that will last you many years. Note that for the delicate electronics soldering you'll most likely be doing, you should use a fine tip on your soldering iron. The majority of soldering irons come with one of these preinstalled, and irons like the XS25 feature an easy-to-replace tip.

I personally use a Tenma 60W Temperature-Adjustable Soldering Station, which retails for around $60.

Stand and Soldering Tip Cleaner

Soldering irons heat up to over 200 degrees Celsius, so it's critically important that you only touch an iron by its handle. I recommend getting a soldering iron stand, which safely stores and holds your iron between solder joints without burning you, your electronics, or your table surface (Figure D-4)!

FIGURE D-4

Soldering stand and sponge

Often a stand will have a built-in soldering tip cleaner. This is used to clean the soldering iron between connections so it continues to work as well as possible. The two most common types of cleaner are a damp sponge or an abrasive pot of brass shavings, shown in Figures D-4 and D-5. I recommend using the brass shavings, as they clean more thoroughly.

FIGURE D-5

Soldering tip cleaner

OPTIONAL EXTRAS

The tools and materials mentioned so far are the basic necessities for soldering, but there are a few other bits and pieces that can make your life easier!

For example, you may want to remove solder from a joint. You can do so with some *desoldering braid* (also known as *wick*) or a *solder sucker*, as shown here. I won't be covering desoldering in this guide, so search online for information and videos showing you how to perform it if you want to improve your soldering skills.

If you're struggling to get solder to flow well, *a liquid flux pen* will help. By applying a special solution called *liquid flux*, the pen helps the solder cling to the solder joint.

You may later find that positioning and holding electronic components while soldering proves difficult. In this case, a *Helping Hands* set, shown here, may be useful.

Often the components that you need to solder together are very small, which can be a problem for those with less-than-perfect eyesight. If this is the case for you, you may find that a magnifying glass (often combined with a Helping Hands) is a vital piece of equipment.

SOLDERING COMPONENTS TOGETHER

Now let's run through the process of soldering two components together. I'll demonstrate by soldering some wire to a motor terminal, which you'll most likely have to do before wiring up the different parts of your robot in Chapter 3.

Preparing to Solder

Before you turn on your soldering iron, first prepare the area. For a soldering space you'll need:

Well-ventilated area   Pick a place that is well ventilated, as soldering produces fumes that you should try not to breathe in.

Suitable work surface   As soldering irons get very hot, ensure you have either a heatproof work surface or some scrap material laid out where you are soldering. A piece of cardboard, a cutting mat, or an old piece of wood will suffice.

Eye protection   During the process of soldering, little bits of solder and flux can sometimes sputter off. I recommend wearing a set of safety glasses or goggles to protect your eyes.

With everything set, place your soldering iron in its stand and then plug it into a power outlet. Wait for your iron to heat up; this may be a matter of seconds if you have an expensive one or a matter of minutes if you have a beginner model like the Antex XS25. Most importantly, remember to *not* touch the tip of your soldering iron while it is plugged in. Even if you unplug your iron, don't touch the tip! It will take a while to cool down.

Once you've plugged the iron in, don't leave your equipment unattended. If you have to leave the room, unplug your soldering iron and wait for it to safely cool down before leaving it. It's not worth taking any unnecessary risks.

Tinning the Tip

Before starting to solder a joint, always *tin* your soldering iron tip. Tinning is the process of coating the tip of your soldering iron in solder. This will make the actual soldering easier and help with proper flow.

First, once your soldering iron is hot, clean it by brushing its tip against your damp sponge or inserting it into your pot of brass shavings.

After this, unwind some of your solder and touch the solder to the tip of your hot iron. The solder should melt on contact. Now cover the bottom quarter-inch of your iron in solder. Use your soldering tip cleaner to clean off any excess solder and repeat this process until the tip of your iron is shiny and covered in solder, like mine in Figure D-6.

FIGURE D-6
Tinning the tip of my
soldering iron

If you are soldering a lot of components, you should re-tin the tip of your soldering iron as appropriate and whenever it's no longer shiny.

Setting Up the Components

It's always important that you prepare the components you're soldering first. Place your parts on your work surface and position them in the way you intend to solder them. For the purpose of soldering some wire to a motor terminal, this involves threading the wire through the terminal, as pictured in Figure D-7. Before you do this, ensure that you have *stripped* the end of the wire first, which means removing about a quarter of an inch of the wire's plastic outer layer in order to expose the conductive metal core. You should use a *wire stripper* for this purpose.

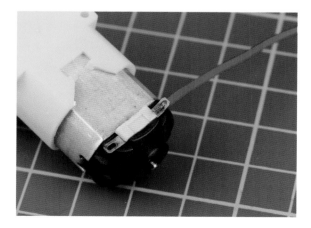

FIGURE D-7
A wire ready to be soldered
to my motor terminal

For other components, you may wish to hold them still using a pair of Helping Hands as mentioned previously.

Soldering the Perfect Joint

Now that we have everything ready, it's time to solder! Spool off about 6 inches of your solder and hold your soldering iron by its handle. It should be in your dominant hand, as if you were holding a pen or pencil. Make sure you're holding the solder wire at least a few inches from the end so you don't burn yourself.

The key to a good soldering joint is *not* to touch the iron to the solder and then melt it *onto* the components. Instead, apply your hot iron to the components you're intending to solder for 2 to 3 seconds to heat the component itself. Then, apply solder directly to the heated components.

Solder flows toward the hottest part of the component, so if you don't preheat the parts you're trying to solder together, you may find that the solder goes onto your iron and forms a messy ball. If this happens, don't worry: just clean your iron and go through the process again.

So, in order to solder a perfect joint, like the one in Figure D-8, follow these steps:

1. Apply the tinned tip of your hot soldering iron to the components you wish to solder together and hold it there for 2 or 3 seconds.

2. With the iron still held against the components, feed the end of your solder into the joint; it should melt upon contact and flow into the joint.

3. When you have melted enough solder that the whole joint is filled, remove the solder but keep your iron on the joint for another second. This allows the solder to flow and settle.

4. Remove your soldering iron from the joint, clean the tip of any excess, and then place it back into its stand.

WARNING

Don't touch the components that you have just soldered right away, as they will still be incredibly hot!

FIGURE D-8

My soldered motor terminal—notice how shiny and cone-shaped the joint is

A good solder joint should be smooth and shiny and resemble a cone, a bit like a tiny volcano. Soldering a wire to a motor terminal is a bit different from soldering a pin to a printed circuit board, for example, so take a look at Figure D-9 to get an idea of what a good PCB soldering connection looks like.

FIGURE D-9

PCB soldering joints on the bottom of a Raspberry Pi (though these have been done by an automated machine, they're still a good lesson)

If your solder joint isn't correct, don't panic! Grab your soldering iron and apply the tip back onto the joint to reheat it. See if the solder flows and settles better, and if not, simply add some more solder to ensure the joint is correct.

Figure D-10 shows a few common soldering mistakes and solutions.

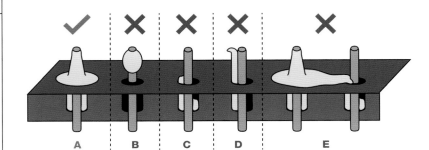

A Solder flows all the way into the joint and forms a smooth volcano-like cone.

B **ERROR:** Solder doesn't flow into the joint and balls up on the pin without reaching the pad.
SOLUTION: Use your iron to reheat the solder, pin, and pad. Add a little more solder if needed.

C **ERROR:** An insufficient amount of solder leads to a weak connection.
SOLUTION: Reheat the joint and apply more solder until you have a volcano-like cone, as shown in option A.

D **ERROR:** Poor connection.
SOLUTION: Reheat and add more solder. Move your iron around the joint to make all the parts hot and to ensure proper flow.

E **ERROR:** Too much solder creates a *jumper*, which is when two pins are wrongly connected.
SOLUTION: Use a desoldering method, such as a desolder wick.

SUMMARY

The ability to solder is a key maker skill that will serve you well—it just requires a bit of practice! Follow the guidance here, or take a look online at some demonstration videos to perfect your technique.

RUN PROGRAM ON STARTUP

THROUGHOUT THIS BOOK, I'VE SHOWN YOU HOW TO RUN YOUR ROBOT'S PROGRAMS USING SIMPLE TERMINAL COMMANDS. WHILE THIS IS GREAT FOR PROTOTYPING, IN THE FUTURE YOU MIGHT WANT TO MAKE YOUR ROBOT RUN A PROGRAM AS SOON AS YOU TURN IT ON, WITHOUT HAVING TO ACCESS IT REMOTELY. YOU CAN DO THIS EASILY BY EDITING THE *RC.LOCAL* CONFIGURATION FILE.

When you power on your Raspberry Pi, it goes through a boot process. When the boot process ends, your Pi looks to the *rc.local* file for any last commands or code to execute. By adding your own custom Python command, you can make any of the programs from this book run on startup. Here's how!

EDITING THE RC.LOCAL FILE

First, make sure the program you want to run on startup is complete and working the way you want it to. It's much better to go through the programming, editing, and debugging process in the terminal rather than wait to see if the program runs when you power on your Pi. This approach will save you from turning your Pi on and off all the time.

Then, from any location in the terminal, open the *rc.local* file using the Nano text editor like so:

```
pi@raspberrypi:~ $ sudo nano /etc/rc.local
```

Make sure you include `sudo` at the start. This command lets you edit with root user privileges so you can save changes you make to the file; otherwise, they'll just disappear!

After you enter in the preceding line, you'll see a file that looks like Listing E-1.

LISTING E-1

The contents of the
rc.local file

```
#!/bin/sh -e
#
# rc.local
#
# This script is executed at the end of each multiuser runlevel.
# Make sure that the script will "exit 0" on success or any other
# value on error.
#
# In order to enable or disable this script just change the
# execution bits.
#
# By default this script does nothing.

# Print the IP address
_IP=$(hostname -I) || true
if [ "$_IP" ]; then
  printf "My IP address is %s\n" "$_IP"
fi
❶
exit 0
```

Now, use your arrow keys to scroll down to the space ❶ between `fi` and `exit 0`. This is where you can add any commands you want your Raspberry Pi to execute on startup. No matter what you add, you must leave `exit 0` unedited at the very bottom of the file.

If you want to run a Python 3 program on startup, insert at ❶ a line that looks like this:

```
python3 /your/file/path/here/filename.py &
```

Replace the filepath with a valid one that points toward the correct directory and your program. Also, make sure to add the ampersand symbol (&) onto the end of the command so that your program doesn't stop your Raspberry Pi from continuing to boot.

After adding the program you want to execute on startup, save your work and exit the Nano text editor by pressing CTRL-X and following the prompts.

A PRACTICE EXAMPLE

Let's say that you want to run the *ball_follower.py* program from Chapter 8 whenever you turn on your robot. To do this, open the *rc.local* file on your Pi and insert this line before the `exit 0` statement:

```
python3 /home/pi/robot/ball_follower.py &
```

Now the last part of the file should look like this:

```
--snip--
# Print the IP address
_IP=$(hostname -I) || true
if [ "$_IP" ]; then
  printf "My IP address is %s\n" "$_IP"
fi

python3 /home/pi/raspirobots/ball_follower.py &

exit 0
```

Let's test it to see if it works. Save the file and then reboot your Raspberry Pi as follows:

```
pi@raspberrypi:~ $ sudo reboot
```

If it's successful, your robot will execute the *ball_follower.py* code. If not, then just remotely access your Pi over SSH and try editing the *rc.local* file again. Make sure that you have the *full* correct filepath and that you haven't made any typos.

That's all there is to making a program run on startup!

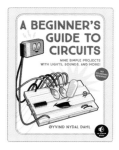

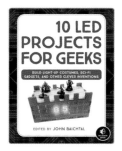